BATTALION COMMANDERS AT WAR

BATTALION COMMANDERS AT WAR

U.S. Army Tactical Leadership in the Mediterranean Theater, 1942–1943

Steven Thomas Barry

UNIVERSITY PRESS OF KANSAS

Published by the University Press of Kansas (Lawrence, Kansas 66045), which was organized by the Kansas Board of Regents and is operated and funded by Emporia State University, Fort Hays State University, Kansas State University, Pittsburg State University, the University of Kansas, and Wichita State University

Library of Congress Cataloging-in-Publication Data

Barry, Steven Thomas.
Battalion commanders at war : U.S. Army tactical leadership in the Mediterranean Theater, 1942–1943 / Steven Thomas Barry.
pages cm. — (Modern war studies)
Includes bibliographical references and index.
ISBN 978-0-7006-1899-6 (cloth : alk. paper)
1. World War, 1939–1945—Campaigns—Mediterranean Region. 2. World War, 1939–1945—Campaigns—Africa, North. 3. United States. Army—Officers—History—20th century. 4. United States. Army—Officers—Training of. 5. Command of troops—History—20th century. 6. Leadership—United States—History—20th century. 7. Tactics—History—20th century. 8. Unified operations (Military science)—History—20th century. I. Title. II. Title: U.S. Army tactical leadership in the Mediterranean Theater, 1942–1943.
D766.B36 2013
940.54'1273—dc23

2013000490

British Library Cataloguing-in-Publication Data is available.

Printed in the United States of America

10 9 8 7 6 5 4 3 2 1

The paper used in this publication is recycled and contains 30 percent postconsumer waste. It is acid free and meets the minimum requirements of the American National Standard for Permanence of Paper for Printed Library Materials Z39.48-1992.

To my best friend and wife,

Mary Ann

Contents

Acknowledgments

I wish to first thank my graduate school adviser, Allan R. Millett, for the intellectual guidance, encouragement, and moral support that helped to make this book possible. I am indebted to him for his infinite patience and support of my career as an army officer and a scholar. I also thank him for his personal interest in my research and leadership within the field of military history.

I also wish to thank Dr. John Guilmartin and Dr. John Brooke for their sound advice, their scholarship, and their service throughout my tenure as one of their graduate students. Over the years, they have served as role models especially in my capacity as an instructor at the United States Military Academy.

Dr. Jonathan House and Dr. Douglas Porch provided me invaluable guidance on how to transform my dissertation into this book. I thank them for their time and sound advice. I know this work is better because of their efforts.

My colleagues from the USMA Department of History and the Ohio State University History Graduate School often provided moral support, encouragement, and ideas that assisted my efforts. Thank you to all of you—there are too many of you to list here.

Also, the U.S. Army afforded me the opportunity to attend graduate school and teach at West Point. I am grateful for these opportunities that many of my peers are not afforded. My experiences as an armor officer for sixteen years, especially in combat, have provided me invaluable perspective in writing this book. I am extremely thankful and honored to serve in an organization that takes such good care of its people. I also wish to thank my primary U.S. Army mentors, Colonels Douglas Cardinale, Richard Coffman, Kevin Farrell, and Clark Lindner, for their endless guidance.

I thank Dr. Richard Sommers and Dr. David Keough of the U.S. Army Military History Institute at Carlisle Barracks, Pennsylvania, for their insightful comments and bibliographic help. In addition, I thank three archivists, Richard Baker, Shaun Kirkpatrick, and Tom Buffenbarger, also of MHI, who showed extraordinary patience and diligence throughout my research trips. MHI is a phenomenal facility that served as the launching point for this work. Also, Susan Lintelmann, United States Military Academy manuscripts curator, spent countless hours combing the special collections on the stuffy fourth floor of the West Point library to provide me critical documents.

Rusty Rafferty at the Combined Arms Research Library in Fort Leavenworth, Kansas, was indispensable for locating key documents regarding 1940–1941 Maneuvers and Operation HUSKY. Dr. Tim Nenninger dedicated his precious time to help locate critical documents in the massive National Archive collections. Without his efforts, I could not have located the information required for a few sections of this book. Dr. Edward "Mac" Coffman provided insightful comments and documents that greatly assisted me in writing chapters 1 and 2. His love of the people who made up the interwar U.S. Army helped to motivate me to write this book. I would also like to thank the Omar M. Bradley Foundation for the two grants they provided me, which funded a good portion of my research trips. Finally, the staffs of the George C. Marshall Foundation Research Library, the Patton Museum at Fort Knox, Kentucky, the Donovan Research Library at Fort Benning, Georgia, Norwich University Archives and Special Collections, the Franklin D. Roosevelt Presidential Library, and the U.S. Army Center of Military History supplied professional and timely help to all of my requests.

I thank my parents, Thomas and Joanna Barry, for teaching me to become the person I am and for giving me their continual support throughout my entire life. They always set the right example for me to follow.

Regardless of any help that I received, I take full responsibility for any errors.

Finally, I could not have done this without the unconditional love and patience provided by my wife, Mary Ann. She is the backbone of our family who keeps the household running smoothly while taking care of three children and a husband obsessed with today's U.S. Army as well as its history. I thank her from the bottom of my heart for her devotion, sacrifice, and encouragement. This book would not exist without her support.

BATTALION COMMANDERS
AT WAR

Introduction

The U.S. Army's junior officer corps—commissioned in the 1930s—proved to be a wise investment as the United States entered ground combat in 1942. The U.S. Army had forsaken robust technological and organizational developments in a budget-constrained decade to maintain quality in its people. Two of these soldiers were Lieutenant Colonels John Corley and George Marshall. Both of these infantry officers were West Point graduates (1938 and 1935 respectively), attended the infantry branch service school in 1940, led units during extensive prewar training maneuvers in 1940–1941, and had never heard a shot fired in anger. As the United States prepared to launch Operation TORCH in November 1942, Corley and Marshall served as typical infantry battalion commanders in the U.S. Army's invasion force. Their backgrounds were similar to the vast majority of their peers. Unfortunately, Marshall became the first commissioned officer killed during Operation TORCH, while leading his battalion's ill-fated assault on the port of Oran, Algeria. Corley led his battalion ashore later in the morning at Oran and continued to serve as a battalion commander until the end of the war. By the end of his career, he had earned two Distinguished Service Crosses, eight Silver Stars, three Bronze Stars with Valor device, and a Purple Heart. He successfully led another battalion in the Korean War and eventually rose to the rank of brigadier general before he retired in 1966. The U.S. Army relied on well-trained regular officers like Corley and Marshall to serve in the key tactical leadership positions as it rapidly incorporated millions of citizens with little or no military experience. Without these officers, the U.S. Army could not have functioned at the tactical level. As the North African and Sicilian campaigns raged across the Mediterranean basin, the U.S. Army's investment in its officer corps produced dividends during America's first engagements against the Axis powers.

This study examines the performance of the U.S. Army officer corps—specifically in the ranks of lieutenant colonel and major—in the Mediterranean theater of operations (MTO) during World War II. This group of officers provided the critical leadership capabilities for an expanded army struggling to execute combat operations in four ways. First, the vast majority of these officers were career professionals before the outbreak of World War II, and their efforts reveal how the U.S. Army made the transition to combined arms operations. Second, this officer corps experienced combat for the first time in the MTO. Third, the historical narrative of the U.S. Army in the Mediterranean provides an instructive model to explain U.S. Army doctrinal, technological, leadership, and training challenges in the early stages of the ground war against the German army, then viewed as the best in the world. Finally, the U.S. Army's rapid mobilization thrust these career professionals into the key tactical leadership and staff positions of battalions and combat commands. This project begins with an analysis of army officer experiences in the 1930s "old" U.S. Army and their roles in the expanding army of wartime mobilization. The fighting in North Africa and Sicily provides the actual combat arena of the officer corps and case studies for analysis and discussion. Ultimately, this study stands as a detailed description of the challenges endured and conquered by a prewar officer corps's attempt to wage modern, combined arms operations in the first stages of World War II.

Most historians use the North African and Sicily campaigns as the departure point to explain how U.S. Army improvement was a linear process throughout the Second World War. The traditional historical narrative of the U.S. Army treats the Mediterranean theater of operations as a deadly training ground for green U.S. forces. Somehow, so the narrative goes, the lessons learned on the beaches of Oran, the hills of the Kasserine Pass area, and the collapse of the Tunis bridgehead all contributed to later success in western Europe. Most authors assume that combat experience is essential for success, and that's just what the units in North Africa needed. However, the U.S. Army did not enter an effectiveness trajectory that neatly began in the Mediterranean. Battalion leaders immediately provided effective combat leadership without any prior combat experience, also known as being "blooded." These first units did not have the time to absorb institutional lessons learned as did the majority of the U.S. Army mobilizing for the cross-channel invasion. Regular Army officers proved to be highly trained, adaptable, and courageous in their first combat experiences in North Africa and Sicily. Their leadership proved to be essential at the battalion level.

Battalions formed the key building blocks of combat commands during World War II. At that level, battalion leaders were center stage at the tactical fight. While the tactical level of war extended all the way to corps headquarters throughout World War II, the regimental, division, and corps headquarters mostly allocated combat resources and coordinated movement of units. Battalion leaders did some of that, but they also closed with and destroyed the enemy on a consistent basis. The U.S. Army lost only three generals to enemy ground combat action in all of World War II.[1] In less than one week of combat in Normandy, the 82nd Airborne Division suffered a 70 percent casualty rate among its battalion commanders.[2] This sampling gives a slight indication of the risk differential at the command levels.

American units throughout the North African and Sicilian campaigns suffered casualties typical of U.S. Army divisions in the European theater throughout World War II. The high-intensity warfare shredded divisions at frightful rates from beginning to end. Combat inflicted roughly the same casualty rates on divisions even after they had endured multiple campaigns. The actions of untested battalion leaders in the MTO did not incur higher casualty numbers as some have suggested. If the U.S. Army proved to be too green in North Africa and an organization that had to take its licks before it got better, then casualty rates should have been much higher. Against some of the most experienced units in the German Army, the U.S. Army battalions that fought from the North African coasts to Messina, Sicily, did not endure higher casualty rates than any other American units in western Europe, 1944–1945.

Up until 31 December 1943, about 5 percent of the officer corps in U.S. Army divisions were regular army. Of that 5 percent, the vast majority of these officers served on regimental, combat command, and division staffs. The Army Ground Forces had an enormous shortage of seasoned officers. To make up the shortage, the U.S. Army had to rely on its officer candidate schools to fill about two-thirds of the officer positions; most of these were lieutenant and captain positions. The company-grade officers who filled these billets would later lead platoons, companies, batteries, and maintenance sections. So, many divisions became top-heavy, with plenty of experience at the top and a significant leadership gap in the actual troop formations. Wisely, the U.S. Army placed regular officers into key leadership billets to compensate for this lack of experience. George C. Marshall, the chief of staff of the U.S. Army, and Lesley J. McNair, the commander of Army Ground Forces, conducted thorough reviews to select the senior officers for divisions, but they did not pick the regimental and battalion cadre officers.[3]

Marshall and McNair recognized that each division needed quality battalion leadership, but they could not hope to scrub every file to ensure a fair distribution. They enforced a cadre system that required older divisions to transfer some of their officers to newly formed units. This system did not result in a truly equal talent distribution. Units that formed first—such as the ones that fought in the MTO—had a higher percentage of quality as well as trained regular army officers and reserve officers. Specifically, the 1st, 3rd, and 9th Infantry Divisions and the 1st and 2nd Armored Divisions escaped the heavier cadre levies due to their earlier deployment overseas and rigorous training program.[4] Prior to their overseas movements, all of these divisions participated in the 1941 maneuver exercises, and the infantry divisions had conducted amphibious training.[5] The divisions that eventually conducted Operations TORCH and HUSKY maintained the bulk of their battalion leaders whom they had trained over the previous eighteen months. The U.S. Army standardized the expansion after it met the operational needs identified in April 1942.[6] All regimental commanders and executive officers were regular army officers. Within each regiment, the three battalion leadership positions—battalion commander, executive officer (XO), and operations officer (S3)—were usually held by an even mix of regular, reserve, and National Guard officers to lead a host of inexperienced and unproven OCS graduates.[7] The MTO battalion leadership in North Africa and Sicily did not follow this typical cadre distribution since they were needed for TORCH. Division leadership retained the best cadre they could before the leveling imposed by Marshall and McNair. The U.S. Army that clambered ashore in Algeria and Morocco had the best crop of battalion leaders to endure the first combat in the European theater. The driving question for this analysis is: how and why did the regular army battalion leadership exercise combat command without any prior combat experience?

This book's thesis is that the battalion leadership exercised by U.S. regular army officers provided the essential component that contributed to battlefield success in the Mediterranean theater of operations despite the deficiencies in equipment, organization, and mobilization and the inadequate operational leadership. Essentially, without the regular army battalion leaders, U.S. units could not have functioned tactically early in the war. The battalion leadership brought discipline, maturity, experience, and the ability to translate common operational guidance into tactical reality. In order to analyze this pivotal group of men, this book will address three sets of factors. First, all of these officers were shaped by prewar experiences. Institutional factors to include professional military education, unit training exercises, and commissioning source formed the foundation of how the U.S. Army prepared these officers for leadership and combat. The second set strives to present a composite picture of the personal factors that shaped these officers. While less tangible than institutional factors, the personal factors include calmness under fire, vigor, and common personality traits. Finally, tactical factors are the "fields of battle" in which the leadership performed their profession. This last category is the most complex and includes the table of organization and equipment of their units, doctrine, the friction of combat, and enemy actions. The friction of combat must be highlighted. In war, nothing is certain, and too many factors exist to allow one factor such as quality battalion leadership to always determine success. The innumerable computations of combat could smile or frown on certain leaders regardless of capability. However, further examination of the three categories above as they played out in combat will show that these battalion-level professional officers were the critical cogs for early Allied success in the MTO.

The legends of World War II—Eisenhower, Bradley, Patton, Montgomery, Rommel—all take center stage in traditional MTO narratives, and the tactical level of war is often obscured by these larger-than-life figures. These generals often take credit for success and pass the responsibility to their subordinates for their failures. Too often, they give themselves front-page billing for how they reformed the army or their respective units. However, a comparison of the tactical level of war with operational and strategic levels is a consistent theme throughout the analysis. The issue of combat experience differs on the levels of war, and these contrasts illuminate the capabilities of the U.S. Army battalion leadership in its first campaigns. An evaluation of combat effectiveness at the tactical level is one of the objectives. This analysis also draws conclusions about the U.S. Army's competence at the three levels of war. The battalion tactical leadership overcame lackluster operational and strategic guidance and other significant handicaps to execute the first MTO campaigns. Furthermore, some historians have concluded that battalion leadership irregularities "seem immeasurable" in the World War II U.S. Army.[8] However, the high quality of the MTO battalion leaders suggests that a more typical group of leaders in later mobilized divisions would not have performed as well to overcome the challenges of the U.S. Army's first campaigns.

Besides addressing the campaigns, this study would be incomplete without a thorough examination of the prewar experiences of these World War II battalion commanders, executive officers, and operations officers. Institutional factors—professional military education, unit training, and leadership development—shaped these men prior to combat. The units that fought in the MTO possessed a high number of regular army officers in battalion command and staff positions. Most of these officers had served on the Pacific frontier on the Philippines Islands, Hawaii, China, or the Panama Canal Zone in company leadership and battalion staff positions. The only non-USMA graduates commissioned in the 1930s were warrant officers, chaplains, doctors, dentists, veterinarians, two hundred Air Corps officers, a handful of enlisted soldiers, and Thomason Act officers—fifty a year—starting in 1936.[9] From 1936 to 1939, USMA graduates still dominated the commissioning at about an 85 percent rate.[10] West Point underwent a training and learning renaissance at the same time that it produced the majority of the new officers. Both USMA graduates and the fifty Thomason Act officers were commissioned in only six branches: engineer, coastal artillery, cavalry, field artillery, infantry, and signal corps. This study remains focused on officers who served in infantry and armor positions chiefly for two reasons. First, these positions traditionally held the key leadership and staff positions in the battalions that conducted combat operations. Second, the line officers had the responsibility to execute combined arms operations. Unlike field artillery officers, infantry and armor officers routinely engaged in direct combat with the enemy and had to synchronize the actions of all branches in combat. The lower casualty rates in artillery battalions gradually made the field artillery branch the most effective of the three (infantry, armor, and field artillery) branches as the war progressed. The technical nature of the branch and its functions coupled with the lower casualty rates produced a very different, but not less important, combat experience than infantry and armor.

Many U.S. officers shared the same "Old Army" skill sets in their early career. Across the U.S. Army in the 1930s, these officers developed familiarity with the systems and doctrine that would prove crucial in the combined arms operations of the Second World War. However, this Old Army inculcated critical components of successful officership in these young men. The 1930s cohort of officers—the future World War II battalion leaders—benefited from the diverse leadership styles of the "Hump" of officers who filled the officer corps in the wake of the Great War. The large postwar influx provided battle-experienced leaders to mentor the 1930s officers. The extremely slow promotion rates for the Hump officers produced both negative and positive role models for new officers.[11] These familiarities provided the leadership foundation that permitted most officers at the tactical level to successfully transition from interwar doctrine to the combined arms operations. A thorough evaluation of the Carolina/Louisiana maneuvers, unit experiences, and leadership challenges such as the Civilian Conservation Corps shows that these served as the key developmental events to explain this transition. The vast majority of battalion commanders and their staff officers participated in some fashion in one or all of these mobilization events. Every battalion commander from the 1st Armored, 1st Infantry, 2nd Armored, and 9th Infantry Divisions participated in at least one of the Army General Headquarters (GHQ) maneuvers.[12]

The U.S. Army's robust investment in its officer corps throughout the 1930s coincided with the Roosevelt Administration's implementation of the New Deal. The New Deal's comprehensive engagement with the average American citizen gradually built confidence that the federal government had the ability to solve complex problems. Furthermore, the extensive programs created a more cosmopolitan society prepared to accept the American government's authority. With the outbreak of World War II, U.S. Army mobilization rapidly absorbed millions of these New Deal citizen soldiers. As they swelled the ranks of units led by regular army officers, their prewar New Deal interactions prevented the creation of substantial barriers between the citizen soldier and the regular. Most importantly, the battalion leaders could foster a command climate conducive to organizational innovation and tactical adaptation. Doctrine and formal lessons learned took too long to disseminate for battalions on the battlefield. Officers who had learned to grasp the fundamentals of combined arms warfare in the 1930s and early 1940s found that the demands of combat required constant adjustment and improvement. Citizen soldiers and their battalion leaders worked together to improve their units' combat effectiveness. In order to do this, battalions had to be learning organizations. The battalion leaders created environments that valued the contributions of their soldiers despite the fact that the vast majority of them had never worn uniforms a year or two before. Their shared belief in the American government's authority prevented severe morale issues that other nations would suffer during the war, such as the British in North Africa.[13]

In order to evaluate the U.S. Army and its officer corps, this study deals with how battalion leaders incorporated technological innovations into combined arms maneuver. Technological concerns normally boil down to the use of Ultra (Allied codename for signal inteligence), tank lethality, command post/radio proficiency and security, and close air support. In technological terms, the U.S. Army had a steep learning curve to overcome in North Africa. Therefore, this book addresses how well U.S. Army battalion leadership employed tank capabilities and close air support doctrine in terms of combined arms operations. Essentially, how did a battalion commander employ the means available to his unit? U.S. Army doctrine and manuals often did not include enough detail or instruction for officers to rely on. Battalions had to develop and refine solutions.

This book also evaluates the competence of the U.S. Army officer battalion leadership in the MTO and dispels the notion that the U.S. Army had to be blooded in order to learn how to fight, mobilize, and train. As with most topics of the Second World War, the historiography is deep. This work does not pretend to be a comprehensive narrative of the Mediterranean theater by any means. The collection of divisions in the Mediterranean provides sufficient diversity for a detailed analysis. The 1st Armored Division and the 1st, 9th, and 34th Infantry Divisions will take center stage in North Africa since they did most of the fighting. The 2nd Armored and 3rd Infantry Divisions had limited action in North Africa, but participated in Operation HUSKY. Finally, the actions of the rangers and elements of the 82nd Airborne Division that participated in North Africa and Sicily will offer detailed examples of elite units. Based on these units, 168 battalion leaders form the starting point for analysis. Each division had approximately thirty battalion leaders in infantry and armor units. Since MTO operations predated the armored division restructure into light and heavy divisions, both 1st and 2nd Armored Division had the same table of organization.[14] Essentially, both armored and infantry divisions had three regiments each, and every armored and infantry regiment had three battalions. Each division also had a reconnaissance squadron that included three more leaders.[15] The study also includes the 509th Parachute Infantry Battalion (North Africa), the 504th and 509th Parachute Infantry Regiments (Sicily), and the 1st Ranger Battalion (North Africa and Sicily) and the 3rd Ranger Battalion (Sicily).[16] Each parachute regiment also had three infantry battalions. The analysis will tend to focus on offensive operations, especially those that include multiple combat arms. The large number of officers and limited primary resources prohibit a discussion of every officer,[17] but a representative selection will present a comprehensive analysis throughout the MTO. This young officer corps developed the sophistication to handle the complexity of combined arms operations at the beginning of the war.

This work contributes to World War II historiography through its analysis of the tactical and operational competence of the Army of the United States. An ongoing debate about U.S. force capabilities rages among military historians, and most tend to depreciate American performance. However, the analysis of battalion leadership in the MTO expands this debate to this oft-neglected group of individuals and shifts focus away from generals, the replacement system, or weapon systems. These individuals had to translate operational guidance into practice despite numerous constraints. Ultimately, this is a "history from the middle."[18] The Mediterranean theater of operations U.S. Army battalion leaders were the middle managers who provided the organizational solution to achieve tactical victories in the United States' first campaigns. Their story explains how these officers mastered combined arms and tactical adaptation at the tactical level to achieve Allied strategic and operational goals.[19]

Although no two officers shared the exact same background or career path, the sum of their diverse experiences developed them all in a similar way—a convergent evolution. Most of the officers developed similar characteristics and capabilities despite different career paths. As army officers matured in similar environments of commissioning source, professional military education, and life in military units, they developed parallel characteristics to adjust to and master their similar environments. This cohort underwent a convergent evolution to adapt to the conditions of the 1930s in preparation for World War II. In preparation for and in actual combat, this evolution occurred at a more rapid pace as time became more limited than experience and resources. The battalion leadership relied heavily on the prewar experiences to adapt on the battlefield. Once there, they worked to ensure their units continually evolved. The adaptability of U.S. Army battalions and their leaders early in the war allowed for maximum agility on the battlefield.

They often used their past experience to develop innovative solutions to battlefield conditions. A well-known example of this tactical diversity is the different solutions to overcoming the hedgerow defenses during the Normandy campaign of 1944. Three divisions—the 29th Infantry, 2nd Armored, and 3rd Armored—all developed separate tactics to overcome the *bocage* despite all being in First Army. Units and commanders surely shared lessons learned, but senior commanders required that subordinate commands develop their own solutions to tactical problems.[20] In North Africa and Sicily, no two leaders implemented their craft in the same manner; their foundation and adaptation in combat produced successful tactics.

As Major General Ernest Harmon watched a battalion of his tanks parade past in Tunis after the fall of Tunisia, he commented: "Then came a battalion of my tanks. They were battle-scarred and dirty, and I was very proud of them."[21] The same could be said of the U.S. Army's battalion leadership. Like the M-4 Shermans they commanded, the battalion leaders entered combat fresh and untouched by the rigors of war. Yet, both men and machines did their job, and without either of them, the Allies could not have achieved their North African victory. Seven chapters tell their story. The first two chapters establish the pre–World War II background of the officers in this study. Further, they describe how similar experiences developed a competent officer corps that filled the critical battalion leadership positions in the MTO. The next five chapters chronicle the combat effectiveness and tactical adaptation of these leaders and their organizations throughout North Africa and Sicily. In this discussion, the combat experiences reveal how the battalion leadership employed doctrine, reacted to the stress of combat, and overcame the challenges induced by an army entering combat for the first time since 1918.

1

West Point Evolution

On 13 June 1933, a cadet captain graduated 177th out of his 346 West Point classmates.[1] He had attained his cadet rank of captain, but other than that, his tenure at the academy seemed unremarkable to his classmates, professors, and tactical officers. After seventy-four days of leave, he arrived at Fort Bliss, Texas, as a newly commissioned second lieutenant in the field artillery. Over the next seven years, he would serve at Fort Riley, Kansas, Fort Lewis, Washington, Fort Hoyle, Maryland, and Fort Bragg, North Carolina. He attended the Field Artillery School at Fort Sill, Oklahoma, for nine months in between his assignments at Fort Bliss and Fort Riley. He held every company level, billet up to battery commander. During the 1940 mobilization and expansion of the U.S. Army, this officer was mentored by a World War I Medal of Honor winner and participated in the Fleet Landing Exercise No. 7 in Puerto Rico, and the Third Army maneuvers in Louisiana.[2] In January 1942, the U.S. Army assigned him as the aide-de-camp to the 34th Infantry Division commander, Major General Russell P. Hartle, and they both left for Northern Ireland to assist with the Operation BOLERO buildup. He was quickly recognized as a quality leader by General Hartle and reassigned to another unit. As the war raged throughout the Mediterranean theater of operations, this officer would lead a battalion, then an infantry regiment, and eventually become an assistant division commander. He earned two Distinguished Service Crosses and a Silver Star, and established one of the most famous combat formations in the U.S. Army and the world: the U.S. Army Rangers. William O. Darby epitomized the ultimate battalion leader while leading the 1st Ranger Battalion in North Africa and Sicily.[3] While not all of his peers achieved as much as Darby, most of them all started their journey at West Point and had similar careers in the 1930s interwar army. Their story begins at the United States Military Academy.

From the National Defense Act of 1920 until 1935, all newly commissioned officers on active duty came from the United States Military Academy (USMA) at West Point. From 1935 until 1940, USMA graduates still accounted for 85 percent of all new officers. In World War II units, graduates prior to 1929 primarily served at the regimental level and higher, due to seniority and age; however, graduates after 1929 mostly served at the battalion level. As the U.S. Army rapidly mobilized from 1940 forward, these regular army officers with a decade or less of service formed the nucleus around which the army congealed. The vast majority of these officers were commissioned from only one source: West Point. The pivotal role of this primary commissioning source requires examination. Their journey as officers began at a time when the academy vastly revitalized its approach to officer development. The academy improved and diversified intellectual development at all levels. Furthermore, it geared its military training more and more to expose cadets to combined arms warfare. The physical fitness program paved the way for young officers to lead and inspire soldiers. Finally, the academy's commitment to bring officers back to the academy after service in field units offered improved role models to the cadets.

After World War I, Douglas MacArthur assumed duties as the superintendent of the United States Military Academy and instituted a major curriculum overhaul. As one historian remarked, paraphrasing the nursery rhyme, "when he was good, he was very, very good and when he was bad, he was horrid."[4] Fortunately, he was good in 1920. His 1920 annual report framed the new vision for the academy to meet the challenges of the post–Great War world. Essentially, MacArthur recognized that West Point was too hidebound and dogmatic.[5] In the future, he claimed, "improvisation will be the watch-word. Such changed conditions will require a modification in type of officer, a type possessing all of the cardinal military virtues as of yore, but possessing an intimate understanding of the mechanics of human feelings, a comprehensive grasp of world and national affairs, and a liberalization of conception which amounts to a change in his psychology of command."[6] The sixty-four-page 1920 report easily eclipsed the fourteen-page average posted by his predecessors and successors. It laid the foundation for the future to develop adaptable leaders prepared for modern war. Like most visions, however, it would take significant time and resources to fully implement. It would also have to overcome 120 years of tradition. MacArthur's wide-ranging reforms alienated many traditionalists and alumni, especially abolishing the summer encampment.[7] General John

J. Pershing, army chief of staff, replaced him and sent him to the Philippines. His successor reversed some MacArthur reforms, but the seeds had taken hold. From 1929 to 1939, the academy had finally fulfilled MacArthur's guidance with a comprehensive curriculum grounded in sound tactical instruction, academic balance, and physical development.[8]

By 1929, the academy recognized its pressing need for land to sustain MacArthur's vision as well as train cadets. Superintendent William R. Smith submitted a request to the War Department to acquire approximately 15,000 acres adjacent to USMA to obtain an adequate water supply, establish an airfield, construct small-arms and field-artillery ranges, and maintain maneuver areas. Seven years and three superintendents removed from MacArthur, Smith maintained that the

> mission of the Military Academy is to train a cadet to think clearly and logically and to do so habitually; to teach him discipline and the basic principles applicable to the various arms in the Military Service; to develop his physique and above all his character; and to teach him to approach all of his problems with an attitude of intellectual honesty, to be sensible of the rights of others, to be inspired by a sense of duty and honor, and unhesitatingly to lay down his life in the service of his country should the occasion arise.[9]

To fulfill this charge, Smith required more land, and he got approval to enlarge the post in 1931. The 71st Congress authorized the purchase of 15,135 acres to expand the military reservation. Besides negating the need to beg for water from the Palisades Interstate Park Commission, the acquisition permitted the academy to conduct quality small-arms, machine-gun, and field-artillery firing as well as maneuvers within ten minutes of the cadet barracks area. A work in progress, the new area required extensive construction and development.[10] The academy slowly acquired chunks of the 15,135 acres as the decade progressed. Tied up in litigation and appropriations, almost all of the land—no doubt accelerated by World War II—finally became part of West Point.[11] The academy utilized the land to train the cadets throughout the 1930s, but it still augmented first-class training with training trips away from the academy until 1939. The academy was expanding its ability to develop cadets while the army successfully fought a 2,000-officer reduction initiated by a budget-slashing House of Representatives.[12] The U.S. Army protected its "training nucleus" and invested heavily in its officer development throughout the 1930s.[13] In 1936, the Congress authorized West Point to increase its class size by about 150 cadets to meet the new 14,000-officer goal.[14] The academy's expansion served as the last key resource to institute the curriculum envisioned in the wake of World War I and the National Defense Act of 1920.

With the dawn of the 1930s, USMA liberalized its instruction. Military training and classroom instruction witnessed the most significant evolutions. General MacArthur's initial sports emphasis stuck, and the academy sustained his emphasis on "the fields of friendly strife" after his departure. Gradually, the academic quality increased. Starting in 1929, the academy began a program to send its humanities instructors abroad on yearlong sabbaticals. For example, the professor of the Department of Economics, Government, and History traveled to New Zealand and Australia to study labor conditions and the results of "socialistic legislation."[15] It also sent instructors, especially those who taught in the sciences, to technical schools or universities in the summer months. Language instructors spent time, some up to a year, in Madrid and Paris, every chemistry instructor attended summer courses at civilian schools, and law instructors went to Columbia University.[16] Also, the USMA library expanded its collection exponentially. Beginning in 1929, the library acquired an average of 1,500 new books annually. Instructor quality improved and approached the standards of civilian institutions such as Harvard. In 1933, West Point defeated Harvard in a head-to-head mathematics competition.[17] A decade previously, West Point had endured harsh criticism for undereducated instructors and cadets.[18] The studies abroad, summer classes, and relevant library not only put those criticisms to rest, but offered an education balanced between the arts and the sciences. Congress's 1933 approval for conferment of bachelor of science degrees to academy graduates validated the reforms.[19]

A New Approach in the Classroom

This scholarship infusion positively impacted the cadets' instruction in every way. The first-class course in military transportation became a practical education course since the academy finally constructed an adequate laboratory with automotive equipment. Furthermore, the class went to Aberdeen Proving Ground, Frankford Arsenal, Midvale Steel Works, Ludlum Steel Company, and Watervliet Arsenal to gain an appreciation for ordnance development and testing.[20] Social science instruction emphasized more lessons in modern history and events. The third-class history course experimented with a Near East regionally focused history course. The third-class cadets also completed a special course in historical research. Military history and engineering course hours totaled 112 hours for the first class, which exceeded the formerly dominant civil engineering course by 8 hours.[21] The academy's academic board even strived to ensure that the entire program reinforced learning objectives. For example, the fourth- and third-class French course textbooks included campaign descriptions of the Battles of Austerlitz and Waterloo to introduce the cadets to military history.[22] Departments devised learning objectives that did not duplicate the work of other departments. The Department of Civil and Mechanical Engineering's military engineering course devoted time to pontoon bridge load calculations, but did not venture into the bridge's construction since the Department of Tactics covered this topic. This detailed lesson planning synchronized a time-constrained curriculum.[23] An analysis of the history program provides an instructive illustration.

West Point slowly adapted its military history curriculum to prepare future officers for combined arms operations. An evolutionary process began after the Great War. The Department of Tactics and the Civil and Military Engineering Department jointly taught a military art class to only the cadet first class. One forty-hour block in the fall term represented the total military history time that first-class cadets received. Three case studies evenly distributed among the class periods drove the thirty-five classes. Surprisingly contemporary, the Russo-Japanese War and World War I led the semester with a heavy focus on Port Arthur and Verdun. Oddly, the Civil War—Grant and the Overland Campaign of 1864 taking center stage—rounded out the course. While a welcomed improvement from the pre–World War I era, the effort remained insufficient to ground the budding young officers. Gradually, the Civil and Military Engineering Department sacrificed some course material and time from its civil engineering and military engineering courses to its military history program.[24] In 1935, for example, the military history course increased by twenty-two periods that included lessons on the Franco-Prussian War, the Russo-Japanese War, and more World War I campaigns.[25]

The Department of Civil and Military Engineering's objective for its military history course aimed to institute a "basic knowledge of military history" that would serve as a foundation for future professional reading and the instruction given at the branch service schools. The department hoped that this basis would provide context to the case studies and vignettes employed at Forts Riley and Benning. Cadets enjoyed—a professor could hope—112 hours of military history instruction throughout their senior year at West Point. Instructors led small-group discussions to emphasize the significance of readings as well as practical lessons learned that the cadets could extrapolate to contemporary operations. Department guidance also stressed that instructors must highlight the personal attributes of the great captains. With texts such as *Great Captains* and *Life of Napoleon*, combat leaders often took center stage. A free-thinking atmosphere that got "the cadets to think" and talk provided the engine for these classes. An operational and strategic levels of war theme acquainted cadets with major campaigns and periods of conflict. The tactical level of war received little attention. The department strived to avoid overlapping instruction with the Department of Tactics. Instead, cadets digested the principles of operations and strategy. They acquainted themselves with combat maneuvers to conduct campaign analysis.[26]

Even while stressing the military history classics and great captains, instructors strove to remain innovative. After two months of Caesar at Pharsalus, Gustavus Adolphus's victory at Breitenfeld, Frederick the Great's defeat of the French at Rossbach, and careful study of Clausewitz and Schlieffen, the military history course spent two weeks on the 1937 Sino-Japanese War. Using restricted official documents, the lesson authors wrote a monograph to present the lessons. This case study employed an interdepartmental approach to illuminate how the Japanese achieved success against their enormous foe. To broaden understanding, cadets not only studied the war's strategy and operations, but also received a lecture on Far Eastern economics. They also received instruction on the Japanese tactics employed in the Chahar and Suiyuan operations—designed to strengthen defense against the Soviet Union—from the Department of Tactics. In each classroom hung an enormous map of China that depicted railroads, major routes, terrain features, and port facilities. By the end of the two weeks, cadets had conducted a thorough campaign analysis. The same interdisciplinary approach based on restricted documents dissected the civil war in Spain. An erudite conclusion on the Spanish civil war lessons emphasized "constant maneuvering" and "motor transportation." The Guadalajara offensive served as the case study to showcase a motorized envelopment to cement this point. Although the conclusion does not rule out the use of the horse in modern war, the analysis revolves around the employment of ground transportation, tanks, artillery, and aviation. Most importantly, instructors stressed that no "secrets of success in modern war" existed for a simple formula.[27]

West Point's intellectual development of cadets kept pace with the branch schools and the Command and General Staff College (CGSC). Both schools incorporated Sino-Japanese and Spanish civil war studies into their curriculum. The branch schools remained fixated on the more practical application to tactics. CGSC included conflict updates and analysis of both wars in almost every quarterly issue of *Military Review* from 1936 to 1940. Typically, these updates varied in length from four to seven pages and presented an operational overview thin on analysis. Actually, the case studies at West Point addressed all aspects of the campaigns more thoroughly than any of the other army schools. Each school did recognize the value in the study of contemporary operations. However, the USMA's rapid integration of current conflicts indicates a sense of urgency among the faculty to prepare cadets for the modern battlefield. It also gave junior officers a common frame of reference with their superiors who studied the same conflicts.[28]

Forging Cadets into Soldiers: Tactical Training

The two charts in this section represent the typical time the Department of Tactics dedicated to tactics, leadership, administration, individual training, and physical fitness. While only one training year, 1933–1934, is used here, the allocations are very similar from 1930 to 1940. The department usually had about 1,500 hours to instruct cadets over their four years. In this example, the cadets had 1,513 hours of training. After the attack on Pearl Harbor, the entire academy allocated more hours to training at the expense of drill, ceremonial procedures, parades, and other administrative tasks. The summer training time almost doubled. Prior to World War II, the academy did not possess the same budget or sense of urgency, but the charts indicate a fair commitment to tactical training. The academy's commitment to about 200 hours of ceremonies seems to be a colossal waste of a future officer's time. The department could have used the time better in the field or on the rifle range. The instructors did try to use this ceremonial time to instill discipline and present the cadets with leadership opportunities. Like it or not, the instructors and cadets were the public face of the U.S. Army. As the pie graph displays, the time spent on field training almost equaled the time spent riding and conducting ceremonies.

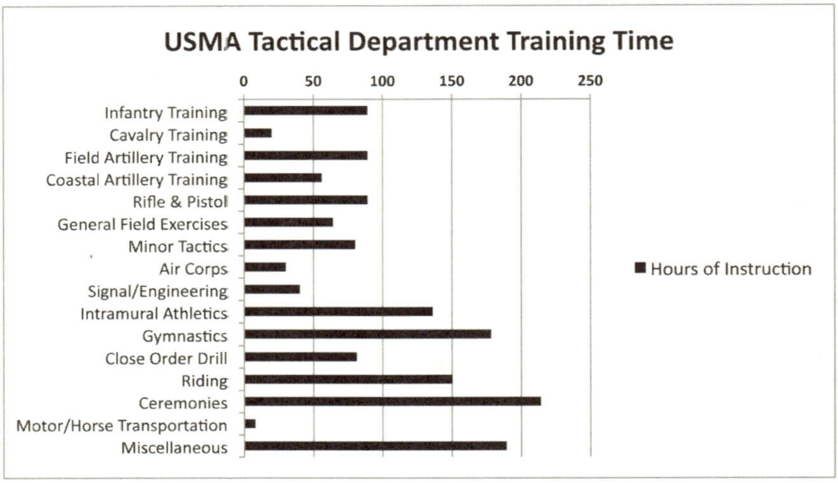

USMA Tactical Department Training Time. This chart represents the amount of hours dedicated to each subject for a cadet over his four years. "Summary of Distribution of Tactical Department Time (Hours)," 1933–1934, Series III, Headquarters, Training Schedules, USMA Special Collections, 8.

West Point never escaped the requirement to maintain nickel-plated tin soldiers on parade until the outbreak of war, but it managed to still focus on preparing proficient cadets.

From 1929 to 1939, West Point executed a comprehensive summer training program for the cadets. The under-three classes spent considerable time on individual tasks such as rifle marksmanship or radio communications. The third class, for example, spent five days on radio communications where they established regimental networks, both wire and wireless.[29] They gradually built on the individual tasks up to platoon-level events such as the "rifle platoon in attack."[30] Of all the classes, the first-class cadets received the most intensive and relevant summer course. Firsties could expect to conduct numerous trips away from the academy. As technologies improved and units deployed to different garrisons, the destinations and duration changed. There were usually two significant deployments. First, the cadets spent approximately a week at an airfield such as Mitchell Airfield, Long Island, and three days at an ordnance post such as Picatinny Arsenal, New Jersey, for technical exposure to flying, maintenance, and ammunition. Second, firsties traveled to Forts Eustis, Monroe, and Benning for hands-

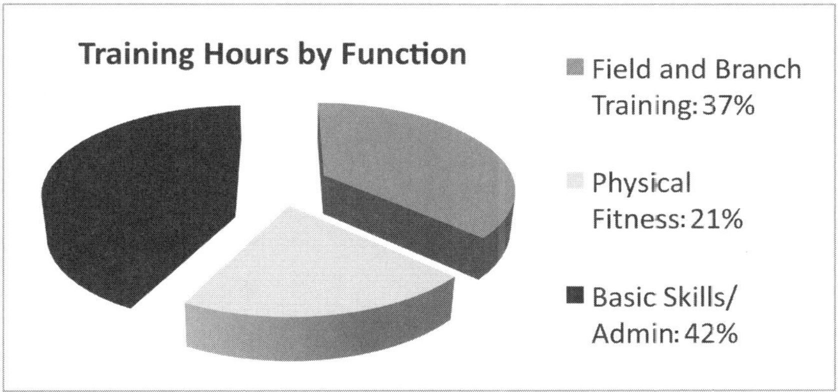

Training Hours by Function. This pie graph delineates the three functions that each training event fell into in the Department of Tactics based on a percentage of time. The department's three-pillar approach strove to develop tactical competency, physical fitness, leadership, discipline, and managerial skills. See the previous bar graph for a subject breakdown. "Summary of Distribution of Tactical Department Time (Hours)," 1933–1934, Series 111, Headquarters, Training Schedules, USMA Special Collections, 8.

on maneuver and weapons training. Instructors and cadets alike lauded the professional development of all the trips.

The weeklong first-class trip to Mitchell Field in 1935 showcases the exposure to the U.S. Army Air Corps. Cadets spent a total of nine and one-half hours on actual flying missions. By the end of the week, each cadet received familiarization and navigation flights and carried out interception and bombing missions. On the missions, the budding pilots operated observation cameras, camera guns while dogfighting, dropped bombs on targets, and had to navigate their air frame. When not flying, cadets received instruction on aircraft logistics, maintenance, and parachutes at the airfield's working facilities. This training was not merely a recruiting stunt to draw officers into the Air Corps. The Air Corps had no problem meeting its quota from West Point. More importantly, the army made a serious investment in every future officer. Any officer worth his salt understood that air power would be an essential element in any future conflict. As such, USMA ensured that all the cadets knew airpower capabilities.[31]

The three-week tactics and weapons trip represented the principal exposure to the army's capabilities. On 13 June 1931, the first class—organized into three battalions—departed West Point for a three-week trip to Fort Monroe, Langley Field, and Fort Eustis. Each battalion was rotated among the posts and devoted five days to training at each location. Langley Field units presented close air support and mechanized force demonstrations and exercises to the cadets. Cadets not only witnessed the tactical employment of tanks, armored cars, artillery, aircraft, and machine guns, but also received detailed instruction on the weapons' capabilities and limitations as well as their use together. The tank section even illustrated the tanks' ability to overcome field obstacles. The time at Fort Meade revolved around ordnance instruction and more tank demonstrations. At Fort Monroe, the battalion divided into sections to fire 8-inch artillery and antiaircraft weapons that included 3-inch guns and .30 caliber and .50 caliber machine guns. The cadet battalions kept score, with 2nd Battalion establishing itself as the clear winner.[32] They even learned how to employ the antiaircraft weapons at night. The chief of coast artillery commented on the cadets' surprising accuracy in a letter to the commandment of cadets. Besides the chief of coast artillery, both the superintendent of USMA and the commandant of cadets observed training at Fort Monroe and Langley Field. The trip even included a side journey to Norfolk Navy Yard, where cadets toured an aircraft carrier, a cruiser, and the battleship *Arizona*. Throughout the trip, cadets participated in athletics from polo to tennis. While many had families in the local area, the officer-in-charge denied all requests for cadets to spend nights away from the posts. The senior officer observation and strict cadet management indicate how serious USMA officers considered this summer trip to be. The academy viewed this trip as the critical exposure to the regular army for cadets.[33]

By the academic year of 1935–1936, this major summer deployment shifted mostly to Fort Benning, with a short stop at Fort Monroe. The shift south reflected the growth of the Infantry School and the slow death of the coastal artillery. Furthermore, the cadets' training had become even more detailed and practical since 1931. In addition to the air training at Mitchell Field, cadets witnessed a whole day of air power at Benning, to include machine-gun strafing, bombing attacks, chemical attack, smoke screens, parachute bombs, air aerobatics, and air-to-air combat courtesy of four different squadrons.[34] Cadets received instruction and conducted practical exercises for ten days at Fort Benning.[35] Demonstrations on the new M1 Garand and antitank weapons exposed cadets to emerging technologies. They also witnessed tanks, mortars, and light and heavy artillery in action. The coup de grace was a battalion combined arms attack.[36] This final demonstration exhibited the synergy of all the nine previous days. By the end of the ten days, the cadets had witnessed the latest tactics and technologies. They also were exposed to the officers, NCOs, and soldiers of the U.S. Army. Many remembered the trip as the most useful experience of their cadet careers.[37] Once the battalions returned to West Point, they ventured into the Hudson Highlands for essential field training.

Back at the academy, they participated in two one-week field problems grounded in field artillery and cavalry tactics at the company and battalion level. As the academy slowly acquired the authorized 15,000 acres throughout the decade, this field training achieved more. The artillery field time included day and night employment of both motor and horse-drawn batteries. The cavalry maneuver offered a weeklong field exercise that culminated in an 80-mile night march. During this maneuver, the cadets rotated positions between rifle, machine-gun, and armored-car platoons.[38] The five-day maneuver included applicable training for future officers. Cadets exercised leadership, conducted tactics, appreciated the difficulty of communications while on the move, and grasped the capabilities of both horse and mechanized cavalry. Instructors produced detailed scripts that outlined the notional combat situation, opposing forces, and learning objectives. The cadets attacked and defended as both mechanized units and horse-mounted cavalry. They even conducted a zone reconnaissance with a rigorous reporting regimen designed to teach the cadets how to use and manage wireless communications. The most important aspect was that the instructors conducted problem critiques—now commonly referred to as after action reports (AARs) or "hotwashes"—immediately upon completion to identify key learning points. The whole exercise incorporated the latest technological developments with an intense introduction to movement and maneuver.[39]

When cadets were not in the field or away from West Point, the Department of Tactics' curriculum utilized the classroom for winter training. From October to April of each academic year, the department instituted a building-block approach to its instruction.[40] The under-two classes were drilled in map reading, combat principles, and squad tactics on sandtables. The second class, nicknamed "cows," had a substantial block of platoon tactics instruction named "minor tactics" and a rigorous course in field artillery. Cows had to solve approximately ten tactical problems. They also had to pass nine written and practical exams on artillery employment.[41] Tactical situations grew more complex for the first class. A combination of terrain exercises, map, and sandtable problems fused the previous three years' training into sophisticated scenarios at the company and battalion level. After observing the local area on terrain walks, cadets were issued maps, a notional combat situation, and a mission. In one example, the general scenario placed the cadets as part of the 2nd Infantry Division that had crossed the Hudson River in the vicinity of Newburgh and had to attack south to seize Highland Falls. As the unit moved south, the instructors fed notional intelligence collected from the air corps and advance guard units. Based on their terrain knowledge, maps, intelligence, and training, each cadet had to solve various situations as part of the 2nd Infantry. In Situation No. 1, "Rifle Company with Machine Guns in Attack," the cadet had to write a very brief operations order of how his company would seize the Central Valley–Highland Falls road junction held by an enemy force with heavy machine guns.[42] The situations varied the unit and weapon types to ensure maximum exposure to all of the combat arms. Instructors then reviewed the students' solutions to lead class discussions about the varied techniques used by the cadets. Beyond the tested situations, classes included extensive lectures and testing on the air corps. The firsties even received command and control training for battalions and regiments. To accomplish this, the department used an innovative command-post demonstration and practical exercise.[43] While the Hudson Highlands winter blasted West Point, cadets continued to hone their tactical skills in their classroom confines. When the snows receded, they carried these lessons forward to their summer training.

The summer training for first-class cadets exposed them to the vast array of career opportunities for a young officer. For most, it was the first time they encountered the "real" army. The trip shaped them all differently, but they had made up their minds about what branch seemed like the right fit.[44] No cadet became an expert in any one branch or weapon system. The cadets, however, did see the combined effects. When they were not in the field practicing their future profession, they reflected on it and thought about tactical problems during their winter curriculum. Given the facility and budget limitations of the day, the academy developed relevant ways to prepare the army's future officers.

Into the Gym and onto the Fields

Physical training and instruction formed one of the key pillars for cadets at West Point. The athletic training cards of James D. Alger, a member of the USMA class of 1935, offer a comprehensive guide. The card cataloged body measurements, physical tests, membership on athletic teams, and section for qualifications. Twice a year, the office of physical education measured each cadet's height, weight, chest (normal, deflated, and inflated), right and left arms and forearms, and right and left thighs and calves.[45] The spring and fall curriculum included biannual grading in gym apparatus, fencing, wrestling, boxing, track and field, swimming, and leadership bearing. At the end of each semester, the cadet received a ranking in his class. Alger's card reflects a 10-pound weight gain, minor increases in chest, arms, and legs measurements, his participation on the corps squad (varsity) track team and the intramural lacrosse and football teams, a standing of 34 for the fall semester and 99 for the spring semester, and a qualification as a Red Cross Life Saver. The office of physical education kept athletic training cards for every year of a cadet's tenure. The daily curriculum included these physical fitness classes, and each semester had dedicated intramural and equestrian periods.

Alger, who would lead a tank battalion at Sidi Bou Zid, Tunisia, and his classmates endured an intense physical fitness evaluation and program. The academy maintained a rigorous intramural program for the fall and spring semesters and throughout the summer training period. The fall semester had a team sport focus that included football, lacrosse, basketball, and track, while the spring semester developed the individual with gymnastics, swimming, cross-country, and pentathlon.[46] If a cadet had mastered swimming and gymnastics then he could participate in team sports like baseball and soccer for the spring semester.[47] The summer training period differed for each class, but the training schedule allowed for both team and individual athletics after military training was complete. To almost a man, cadets thoroughly enjoyed the intramurals. They thrived on the competition and viewed the games as a tremendous release of the pressure of academics and military training.[48] More importantly, it prepared them to lead, as one officer remembered:

> I had the thought [that] the intramural program was terribly important because as quickly as the officers went to their regiments, they had to coach, referee, and deal with problems of athletics and sports in their outfits. So, I thought it was really a great thing to prepare the young lieutenant to the point where he could deal with the men out of high school, or out of college that were in the enlisted ranks and understand them better. If he was

> nothing but a bookworm, he would have trouble getting through to some of these people.[49]

Besides inspiring most to a lifetime of physical fitness, this program encouraged physical toughness, leadership, teamwork, and endurance.[50]

The USMA emphasis on physical fitness prepared future officers for service in an army committed to athletics. The first Army-Navy football game—1 December 1890—ushered in a mandatory physical fitness program at the academy. MacArthur's post–World War I reforms made athletics and intramurals mandatory for every cadet. This acceptance of athletics as a fundamental tenet to cadet development rapidly bled over to the whole army. To abolish high incidences of drunken behavior, venereal disease, and, above all else, desertion, officers implemented extensive sport and athletic programs. The World War I trench warfare confirmed the importance of physical fitness to survival on the battlefield. So, while competitive sports like baseball and volleyball continued to expand, every soldier practiced combatives, boxing, jumping, and vaulting. Officers and soldiers embraced organized athletics and physical fitness as proper training for combat, a way to build camaraderie and cohesion, and a critical component of leadership development.[51]

Sports became even more important to units in the interwar period, especially for those guarding America's Pacific possessions and the Panama Canal. The officer corps played polo, golf, and tennis in the afternoons after they finished their official duties in places such as Fort Stotensburg and Fort McKinley, Philippines. Officers—mostly cavalrymen—even credited polo as the catalyst for officer retention in the interwar years as well as the development of aggressive and highly adaptable officers.[52] In an era of brutal budget constraints, the officer corps found professional development, discipline, and fun in organized athletics.[53] The Sports Carnival in the Philippine Islands represented the high-water mark. This three- to four-day event combined every conceivable sporting event with social activities to determine the best athletes and, of course, socialize.[54] More importantly, the U.S. Army's attachment to athletics demanded officers to maintain appropriate levels of physical fitness throughout their careers. This created proper role models for enlisted soldiers and noncommissioned officers to follow.[55]

Role Models

Constant exposure to army officers offered the most profound leadership lessons for cadets. The Martin "Marty" Mahers were few and far between at West Point.[56] Very few civilian instructors, especially outside of the athletic department, were NCOs or civilians like Maher. Cadets encountered officers of all types throughout the entire day. Through their interaction with officers, cadets took the full measure of the weak, mediocre, and outstanding. From the classroom to the cadet barracks to summer training, the cadets absorbed the officers' leadership.[57] Regardless of the leadership quality, cadets shaped their concepts of what makes a good leader by seeing many in a short time period. Per academic year, cadets usually took six to seven classes and had, at a minimum, twelve to fourteen different instructors. The actual number was usually much higher since academic subjects often "resectioned" after the first thirty days to place cadets of equal ability in the same classrooms. When the course resectioned, cadets received another instructor.[58] Beyond the classroom, each cadet had a company tactical officer. Many company tactical officers doubled as Department of Tactics instructors. So, throughout the summer and winter tactics curriculum, most cadets saw every company tactical officer as well as the branch detachments that augmented this training. Academy officers also served as coaches and assistants for athletic teams, offering another exposure outlet. The rapid turnover of officers—each department usually rotated two to four officers a year back to the army—contributed further to a high turnover rate. After only two years, for example, John Waters replaced Omar Bradley as a tactical officer. Some officers had more impact than others on cadets based on time, experience, and personal identification. A few cadets did not "recall anyone that was inspiring," but they remembered being "impressed" with certain officers.[59] A conservative estimate determines that the average cadet had significant contact with at least sixty officers over his four-year tenure.[60]

As is often the case in leadership, the negative leaders shine brighter than the positive ones. West Point proved no different. In one incident, a cadet detested the cavalry branch from his one interaction with his riding instructor, a cavalry officer. The instructor—a Captain John Fryerson—exhibited unbridled cruelty toward the riding horses in front of the cadets. He consistently abused the horses, even whipping them while someone held the reins tight. The cadets uniformly hated this man and by association, his service branch. The academy, however, assigned two young cavalry officers (one of them was John Waters) the next year to eradicate the negative cavalry branch image. Both officers quickly gained reputations as tough but fair and impressive leaders. The cadets thirsted for quality role models, and the younger officers could provide constructive examples of how to lead and not to lead.[61]

The three charts in this section characterize the USMA officer composition throughout the 1930s. To ensure an even sampling, three years—1930, 1934, 1938—were selected to represent the officer distribution for the entire decade. The three years are typical when compared to the other years. The data are extracted for seven different departments to give a sense of what type officer a cadet would encounter throughout his four-year career. The branch detachments, U.S. Corps of Cadets tactical officers, and Department of Tactics conducted all gymnastics, winter and summer training, and disciplined the cadets throughout the year. Regardless of academic year, every cadet encountered these officers on a daily basis. History, modern languages, mathematics, and civil and military engineering included subjects taken by every class. The academic courses selected relate not only the curriculum breadth, but they represent the subjects taken by all four classes at any time. While no fourth-class cadet studied engineering, for example, since it was mostly a first-class subject, all fourth-class cadets took mathematics and French. All the instructors hailed from nine branches: infantry, field artillery, cavalry, engineers, coastal artillery, air corps, signal corps, quartermaster, and ordnance. The branch type and vast distribution indicate the cadet exposure to every officer type. Infantry, field artillery, cavalry, and engineers tended to dominate this population, while air corps, signal corps, ordnance, and quartermaster officers were a small minority. The charts do not reflect the rank, but approximately 60 percent of the officers were first lieutenants and 30 percent were captains.[62]

These data indicate that cadets interacted mostly with combat arms officers with less than fourteen years of service. The constant exposure to role models ingrained positive leadership traits and repudiated negative qualities. Academics and the longing for graduation could consume cadets.[63] The frantic pace often prevented them from forming lasting mentorship bonds beyond one or two officers. Wherever they went, officers stood ready to instruct and evaluate them: in the riding hall, during a sandtable exercise, or while reciting French. The myriad of officer-cadet combinations eludes a definitive quantitative or qualitative role model analysis. The data do suggest that cadets had ample opportunity to form opinions about their own leadership style. Most found the cadet leadership wanting, silly, and cruel, so many looked to

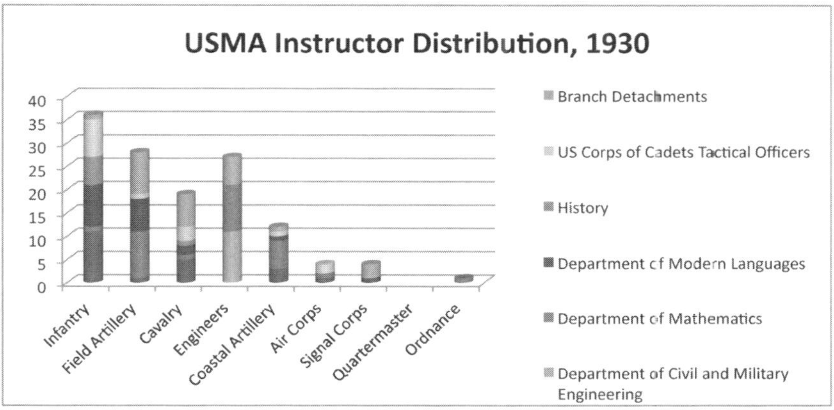

1930 USMA Officer Distributions. All data compiled from Official Register of the Officers and Cadets, United States Military Academy, years 1930, 1934, 1938, USMA Special Collections. To ensure fair sampling, the same data were compared with years 1929, 1931–1933, 1935–1937, and 1939–1940. The results mirrored the three years displayed here.

the officers for inspiration.[64] When the cadets finally graduated, they had seen a vast array of officers to form their opinions about what type of leader they wanted to be.

By the 1930s, USMA had settled into the practice of bringing officers from the army back to West Point to instruct. Prior to the MacArthur reforms and the academy's slow and reluctant adoption of his changes throughout the 1920s, the school had relied on retaining newly graduated second lieutenants to teach subjects they excelled in during their cadet tenure. This practice had many drawbacks, including immaturity, limited scholarship outside of the academy, and, critically, no real ability for officers with practical experience to serve as role models for the cadets.

A few future MTO battalion leaders such as John Inskeep and John Waters returned to the academy after their first couple of assignments. Inskeep taught math to fourth- and third-class cadets, and Waters served as a tactical officer. A 1931 graduate, Waters returned to West Point in 1936. The commandant initially turned down Waters's request for assignment since there were no vacancies, but a month after the initial request, Waters's request was approved. He served a year as a transportation officer and three years as A Company's tactical officer. As a tactical officer, Waters assumed complete responsibility for the disci-

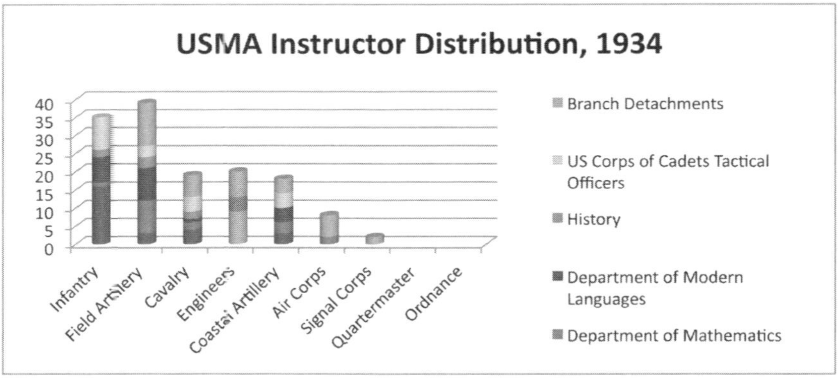

1934 USMA Officer Distributions. All data compiled from Official Register of the Officers and Cadets, United States Military Academy, years 1930, 1934, 1938, USMA Special Collections. To ensure fair sampling, the same data were compared with years 1929, 1931–1933, 1935–1937, and 1939–1940. The results mirrored the three years displayed here.

pline, training, and leadership of 90 to 100 cadets. A 1931 graduate and a tactical officer from 1937 to 1940, Waters witnessed the academy's continued evolution. The increased emphasis on mechanization and adoption of new weapons forced Waters to grasp these new concepts since he had to teach the new tactics and procedures to cadets almost every day. From the more ceremonial duties at his first Fort Meyer post to the Cavalry School and now to West Point, Waters continually improved his understanding of combined arms warfare and watched as the regular army did the same. During a 300 air corps fly-by and the armored force tour at West Point, Waters recollected, "They (the cadets) began to get a feel, as I recall then, that this Army is changing now. Look at what we are going to get. Look at those airplanes up there. Look at those vehicles."[65] As reluctant as any cavalrymen to give up his horses—he owned six—Waters grasped the mechanization and combined arms concepts. The spread of global conflict infected the instructors and students with an attitude of "seriousness" when they tackled military training.[66] Waters simultaneously developed future leaders and himself. He serves as one example of the many role models for the cadets on the banks of the Hudson in the 1930s.[67]

The typical junior officer before World War I engaged in stability operations across the Pacific empire. Army units rarely conducted major combat operations in any form. West Point recognized this

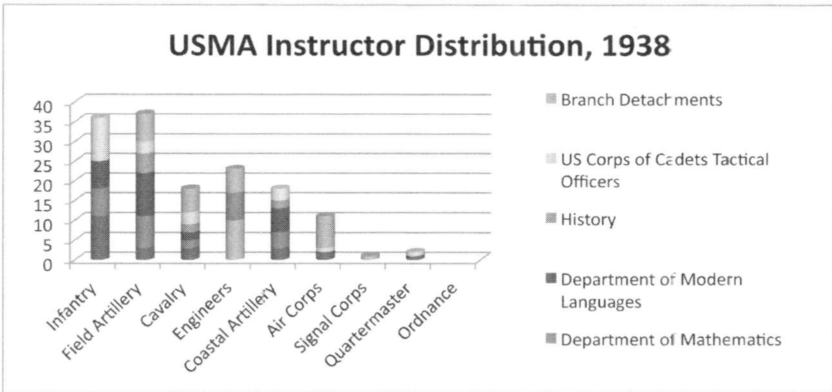

1938 USMA Officer Distributions. All data compiled from Official Register of the Officers and Cadets, United States Military Academy, years 1930, 1934, 1938, USMA Special Collections. To ensure fair sampling, the same data were compared with years 1929, 1931–1933, 1935–1937, and 1939–1940. The results mirrored the three years displayed here.

dilemma in the cadets' professional development. The cadets' intense study of general war proved to be a way to introduce them to high-intensity experiences. Course study at the branch schools as well as the Command and General Staff College at Fort Leavenworth continually built on this foundation. As extreme ideologies permeated the globe, officers recognized that a combat experience chasm had developed in its company-grade officers. Aggressive advances in Africa, China, and Europe in the 1930s raised suspicions of a major conflict in the future. A shift in focus to educate cadets less in the science of war and more in the art signaled a strong break with the academy's traditional math and science curriculum.

Overall, the cadets' time at West Point proved essential to their professional development. The academic renaissance dramatically improved the classroom instruction, the curriculum, and the instructor quality. Cadets graduated with grounding in languages, engineering, history, philosophy, law, mathematics, and economics.[68] Physically, the program developed leadership qualities in the cadets that prepared them to lead athletics in their future units. The military training did suffer at times from resource constraints and distractions that included excessive emphasis on close-order drill and ceremony.[69] Beyond these limitations, the training, in and out of the classroom, built the foundation on which cadets could expound in their future assignments and duties. The

Departments of Tactics and Civil and Military Engineering incorporated the latest developments into the instruction whenever possible. The constant contact with role models—both positive and negative—provided cadets an opportunity to develop their leadership qualities. Whatever the shortcomings, the cadets witnessed the changes in modern warfare. While they were not totally prepared to execute and endure all of its challenges, their minds were sown with the potential.

2

Forging Leaders beyond the Hudson

Like the rest of the U.S. Army defending the Bataan Peninsula, Major Floyd F. Forte awaited the inevitable outcome of Japanese victory in the spring of 1942. Forte decided to pen a last letter to his friends, Major and Mrs. John W. Merrill. He did not even have their correct address and sent the letter care of the U.S. Army Adjutant General in Washington, D.C. The letter—dated 13 April 1942—recounts "Sammy" Forte's experience as a company commander, a regimental commander, and finally a corps staff officer as well as his impressions of the Japanese. Besides these professional summaries and his fond memories of life prior to the Japanese attack, Forte dedicated half of this last letter to two important thoughts. First, he addressed the reformation of the officer corps: "John, old man, remember how we use to sit around drinking and bitching about the deadwood in the army. I believe you'll have a chance to do something about it. For God's sake get rid of it. We can't stand for incompetence or inefficiency in our Army. We can't let sympathy for social friends or their wives prevent our kicking the unfits out of the Army."[1]

His second point confirmed the value of his professional education at West Point: "It's been a wonderful life. If there's no more remember Sammy has no regrets. I only hope one thing above all and that is that I'll live up to our song 'Alma Mater.' I was thinking about that once when I was in a particularly tight spot, it's really remarkable how appropriate it is."[2] The young major's reflections grasped two components essential to the officer profession: standards of performance and adherence to duty, honor, and country. Unfortunately, Sammy Forte could only impart this wisdom and not practice it after the fall of the Philippines.

The U.S. Army reported Major Floyd "Sammy" Forte missing in action (MIA) on 7 May 1942, and the War Department listed his official date of death as a POW on 31 July 1944.[3] In his last letter before he was MIA, Forte reached out to a dear friend not only to reminisce about their friendship and profession, but to proffer advice on the future of the U.S. Army at the outset of World War II. He did not stand out as remarkable in the USMA class of 1934. He graduated 200 of 251 and never scored higher than 115 (in Demerits) in ten measurable categories tracked by *Official Register of the Officers and Cadets*.[4] By all accounts, West Point classified him as below average compared to his peers.

Commissioned an infantry officer, Forte had a career that followed the typical trajectory of a young officer. He attended the infantry school, regular course, 1938, and deployed to the Philippines, as almost all officers did at some time in their career. In this short career, he commanded a company and regiment in combat and mentored one of his lieutenants to take over his company. This lieutenant, Alexander R. Nininger, received the first Medal of Honor of World War II for the U.S. Army while leading Forte's former company in January 1942 in the defense of the Abucay Line.[5] Forte's inauspicious beginnings at West Point and untimely demise belie the importance of officers like him who served in the rapid mobilization of the U.S. Army in World War II. The regular army officers—year groups 1930 to 1940—formed the tactical backbone of the U.S. Army formations that seized the shores of North Africa and Sicily. Of the 267 USMA graduates left to defend the Philippines, 173 officers like Forte never left the islands alive. Fortunately, the U.S. Army did not squander the majority of these officers in the hopeless defense of the Philippine Islands.[6]

Forte's background bridges the gap between West Point and the two issues of professional military education and unit experience. Tested before his MTO colleagues, Forte could call on prewar experiences that prepared him for the rigors of combat. Prior to World War II, company-grade officers like Forte shared common duties that primed them for combat. First, they all served in a line unit of their branch in the United States and usually a two-year tour in the Pacific. In their first units, the young men were inundated with new duties and responsibilities. They learned how the army "worked," and they plied their trade for the first time in garrison and the field on maneuvers. Second, unsuspected development opportunities sprang up that broadened their leadership competencies. The creation of the Civilian Conservation Corps touched about 25 percent of the officers. Also, the slow officer corps expansion begun in 1936 forced the 1930s officers to develop and intermingle with Thomason Act officers. Third, the branch schools gave the officers a nine-month period to increase their professional qualifications. Finally, the large-scale army maneuvers—to include the General Headquarters maneuvers of 1941—seared tactical lessons learned into the minds of impressionable young officers operating at the battalion level.

Regular Army Training

Whether serving around the Pacific Rim, along the Panama Canal Zone, or at the myriad posts in the United States from Fort Ord, California, to Fort Myer, Virginia, newly commissioned officers came to know how the U.S. Army ran in the 1930s. At any one time, approximately one-sixth of the officer corps served in overseas garrisons and participated in the most robustly resourced annual maneuvers, especially in Hawaii and the Philippine Islands. Officers stationed inside the continental United States often had to stretch their training resources a little further during smaller maneuver exercises. Regardless of the duty location, junior officers served in line infantry and cavalry units. While serving on the line, they gained a more complete understanding of their responsibilities as officers and gradually gained more combined arms maneuver experience with each maneuver exercise.

When new lieutenants arrived at their new duty stations straight out of West Point, a mountain of duties overwhelmed them. From Fort Devens, Massachusetts, to Fort Bliss, Texas, senior officers expected their new charges to grasp as many army functions as possible. A seemingly endless list of additional duties plagued the young officers: counting canned goods in the post exchange, movie officer, gymnasium officer, court duty for the defender and prosecutor, supply officer, stable officer, mess officer, and transportation officer to name a few representative examples.[7] To ensure a fair distribution, the duties often rotated on a biannual basis among a company's officers. Not only did these additional duties rotate between officers, but so did duty positions. For example, infantry officers could expect to be rifle platoon, cannon platoon, and machine-gun platoon leaders. Service for combat arms officers even included time in service companies, where they often learned about transportation, maintenance, and convoy operations. One officer did only a two-month stint in a service company, but remembered that his job of running a convoy for the St. Patrick's Day Parade in Boston was a critical learning event in his career.[8] The development did not end with the rotation of additional duties and assignments.

Senior officers and noncommissioned officers ensured that their junior officers became technically proficient on the units' weapons. Fortunately, much World War I knowledge still permeated the "Hump" officers and their NCOs. They knew how to train soldiers. They also knew the value of trained junior officers. NCOs spent afternoons teaching their new lieutenants to lay in a mortar with an aiming circle, put a mortar into action, and pack machine-gun barrels to avoid stoppages. On the range, senior officers and NCOs taught the lieutenants how to maintain the machine guns and find ways to maximize resources. In an era of furloughs and constrained budgets, units had to search for innovative ways to make every resource count. For example, they developed "one-shot" drills to qualify everyone on a machine gun. Using one round, the officer shot into the first square of a qualification target. Next, the officer went through the whole search and traverse sequence without any more ammunition. Also, company, battalion, and regimental commanders conducted terrain exercises in classrooms and on the actual ground to explain tactics to their junior officers. Certainly, weapon systems and tactics constantly change, and these skills atrophied. But these efforts imparted the knowledge to junior officers of how to train individual soldiers and junior officers. They also grasped that training never occurs in an unlimited resource environment.[9] Officers worked long hours to train themselves and their subordinates, and many a family did not remember seeing their fathers around much during the typical workweek.[10]

Annual maneuvers developed problem-solving capabilities in junior officers. For example, interwar military maneuvers in the United States required units to deploy to larger army posts and training areas. This movement of forces—not technically training—ensured officers had to go through the machinations of actually moving forces and equipment. Hamilton Howze's unit movement from Fort Oglethorpe to Fort Benning represents a typical weeklong trip. While the majority of the 6th Cavalry went south to Benning via motor convoy, the unit had to move all of its equipment and horses by train. The commander charged First Lieutenant Howze and eighteen men to care for and load 600 horses. Howze and his overworked men accomplished the deed, although they arrived exhausted for the maneuvers at Fort Benning. Certainly the army did not rely on the movement of horses to fight World War II, but it did rely on the problem-solving skills that situations like this developed in its officer corps. Howze never forgot the adversity of this task.[11]

Horse wrangling represented only one positive leadership aspect for officers during continental United States maneuvers. Once most regular army officers handed off their CCC duties to the reserves, the U.S. Army held annual maneuvers for its units on a regular basis in the mid-1930s. Many officers remembered that the training was not always resourced properly, but the units took the training seriously. They wanted to perform well for the major training event of the year. To prepare for regiment-versus-regiment maneuvers, regimental headquarters conducted situational training lanes for companies and battalions months prior. Jim Gavin remembered intense training exercises at Fort Ord, at both the regimental and division levels throughout the entire year, including amphibious exercises.[12] Among the junior officers, innovative ideas permeated the maneuvers to achieve maximum effect. One example highlights these efforts to master combined arms warfare. During maneuvers at Pole Mountain, Wyoming, frustration mounted as ground units could not talk to aircraft supporting the maneuvers. A motivated lieutenant, Jean Engler, liaisoned with the pilots to learn which radios he needed to speak with them. He then converted one of the new reconnaissance cars into an ad hoc command post that coordinated ground and air reconnaissance. Engler augmented this integration by actually flying with the pilots to appreciate the difference between air and ground observation. Once his commander saw this modification, he ordered Engler to take it one step further. They developed the reconnaissance truck into a mobile command post complete with radios, command desk, map display boards, and supplies. As the unit became more and more motorized, the scope and pace of their tactical world increased exponentially. Company-grade officers knew that they had to integrate new technologies and techniques to adjust. The mobile command post was just one of many examples that revealed their adaptation.[13]

Other large exercises tested the mechanized organizations that the army began to develop in the late 1930s. The 13th Cavalry exercises of 1938 tested the mobility and organization of the mechanized brigade for just over a month. The brigade road marched its entire formation from Fort Knox, Kentucky, to Fort Riley to execute five different combat scenarios and then road march back to Fort Knox. This arduous event exposed its officers to the challenges of maintenance and logistics of a mechanized formation as well as the speed of its employment in tactical scenarios.[14]

The annual maneuvers served as the capstone training events for company-grade officers. Commanders structured a gradual training glide path prior to the annual maneuvers. At their local training areas, units conducted maneuvers at the squad, platoon, company, and battalion levels. These smaller collective events gradually prepared new platoon leaders and company commanders. So, junior officers gained firsthand appreciation of what the army termed "training management."[15] Training management outlined how officers should design training schedules and resources to produce a combat-ready unit. The battalion-level leaders of the U.S. Army units that stormed ashore in North Africa and Sicily often had to implement crash-course training management schemes. These battalion leaders simply recalled how they had built their unit's capabilities as company-grade officers. While training management seems intuitive, the true challenge is obvious: can an officer execute a training plan under severe time constraints? Tough choices have to be made on what training to cut without creating a gap in capacity. Young officers attained this quality judgment from the dogged mentorship of senior officers.[16]

This continued in overseas units and often provided opportunities for officers to execute orders with little to no oversight from commanders. The U.S. Army, for example, practiced maneuvers in Panama to defend the canal. The severely compartmentalized terrain, jungles, and swamps of the isthmus did not mirror many Mediterranean battlefields, but the maneuvers did have professional value. Leaders had to execute orders with very little supervision or oversight from their higher headquarters. The two-week exercise engaged the entire Canal Zone and proved very "strenuous, comprehensive, and instructive" for peacetime maneuvers.[17] Company-grade officers often executed missions based on their own initiative and analysis.[18] Austere jungle conditions and the tropical environment required leadership by example in the maneuver units. Soldiers and officers alike endured hardships on maneuvers that tested their mettle and made them tough.[19] If nothing else, officers learned how to fulfill their commander's intent. This skill proved critical during mobilization training and, more importantly, in battle when units operated from oral or very abbreviated operations orders.

Many officers requested overseas tours in the Philippines, China,[20] or Hawaii for the exotic locations as well as the professional value of these assignments. Every officer who served in the Philippines realized the strategic futility of defending the islands and knew they were a "gone goose" if and when Japan attacked.[21] The real value of the training rested at the tactical level. The heat and rains were oppressive and hampered quality training at times in the late spring and summer; however, the Philippine units maintained high professional standards that kept their regular officers trained.[22] The U.S. Army conducted maneuvers every year to practice the defense plans and challenge the units. Units moved mostly at night over vast distances in choking dust and intense tropical heat. During a maneuver to defend Lingayen Beach, an entire regiment's officers pleaded with the commander to halt the march since the heat had inflicted casualties among the soldiers and pack animals, to include the death of a few mules. Ignoring the calls for a halt, the regiment continued and stopped to establish its assembly area that night. The colonel castigated the entire regiment's officers for their lack of fortitude. He wondered how the unit would react if they had to contend with the enemy and the weather conditions. His chastisement created a positive attitude that pervaded the whole regiment, which pushed on to the objective even when the weather worsened over the next few days.[23] These field exercises simulating war transmitted leadership lessons and experiences to junior officers.

Lieutenant Edwin Van Valkenburg Sutherland's training experience mirrored those of his peers on the Pacific frontier. He served at Fort McKinley, Philippine Islands (known as PI), from 1937 to 1939. Known to his peers as "Van," he commanded Company F of the 57th Infantry Regiment (Filipino Scouts). In anticipation of a Japanese attack, the Philippine Division conducted annual maneuvers to rehearse different defense scenarios. In 1938, the division, along with Sutherland's Company F, maneuvered southward along the length of the Bataan Peninsula to practice a withdrawal under enemy pressure. This prescient scenario forced the units to occupy successive lines of resistance until the final defensive belt on the peninsula's tip that offered a clear view of the "Rock"—Corregidor island. Throughout these maneuvers, Lieutenant Sutherland led his ninety Filipino Scouts and communicated with his regimental headquarters through wireless radio exclusively. Although he likened himself and his fellow infantrymen to Neanderthals compared to signal corps officers, he operated in an environment that required innovation and critical thinking.[24] Unlike his doomed comrades four years later, he and his company ate well—to include monkey meat—but only through constant coordination with the regimental supply trains. This prewar army did not enjoy a time of plenty when it came to quality equipment. For example, the units still relied on World War I ambulances for official transportation. Annual exercises constrained by frugal training budgets challenged young officers. Like Sutherland, they often had to lead in a harsh tropical climate. Company-grade officers had to understand the division's role in Plan ORANGE and got to see for themselves the movement and synchronization of infantry, artillery, cavalry, engineer, and medical regiments.[25] These exercises made indelible impressions on officers at a young age. The maneuvers challenged and informed them as to how the U.S. Army operated beyond the small world they usually knew in garrison life.[26]

Of all the Pacific outposts, Hawaii conducted the largest exercises with the most troops. Almost a tenth of the army's regular officers were stationed in the Hawaiian Department at any one time.[27] The Hawaiian Division executed brigade and division-sized maneuvers for three to four months to rehearse the defense of Oahu against a Japanese assault. A joint exercise with the U.S. Navy, known as Grand Joint Exercises, served as the final capstone event for the year's training. Besides the habitual maneuvers and unit training such as machine-gun practice on the beaches, the department produced innovative training events.[28] Whenever possible, they encouraged combined arms training, air-ground cooperation, and communications and command-post exercises. Officers leased additional land for the training and pleaded for more and more resources.[29] The department commander performed annual tactical inspections of not only the large-unit maneuvers, but also small-unit combat exercises and day-night machine-gun firing and unit movements.[30] The detailed inspections recognized training and resource shortcomings such as machine-gun ammunition and motorization shortages. Company-grade officers serving in Hawaii spent significant time in the field and witnessed maneuvers up to the brigade level routinely. They were exposed to joint and combined operations in a resource-constrained environment. Nonetheless, their battalion commanders worked hard to educate them in their trade, as one commander remembered during a joint exercise: "The training was repeated frequently so that its execution would become instinctive in case the operation had to be conducted during hours of darkness."[31] Mentally, the time in Hawaii shaped their approach to warfare. Certainly, the benefit of the Plan ORANGE maneuvers in Panama, PI, and Hawaii for young, impressionable officers seems obvious, but what other factors shaped their leadership development?

The Civilian Conservation Corps and New Deal Impacts

The army's glacial promotion rate increased junior officers' contact time with troops. Thus, officers new to the U.S. Army spent their formidable company-grade careers learning their trade on the job. Hamilton Howze arrived at his third duty assignment as a newly promoted first lieutenant after five and a half years as a second lieutenant. By comparison, an army officer who graduated West Point in 2009 as a second lieutenant is promoted to first lieutenant in eighteen months. Serving in the 6th Cavalry at Fort Olgethorpe, Howze inspected Civilian Conservation Corps Camps and participated in several annual maneuvers at Fort Benning. John K. Waters received quality mentorship from his troop commander—a World War I officer who remained on active duty—while he served at Fort Myer in Washington, D.C. While Waters was at Fort Myer, Lucian Truscott, usually recognized as one of the best World War II division commanders, and possibly the best, also mentored him. As a young officer, Waters attended III Corps maneuvers at Fort Meade and maneuvered as a squadron in the vicinity of Fort Belvoir. However, Waters participated in the U.S. Army's nadir. The eviction of the Bonus Marchers from the capital served as his first "action." In the wake of the burnt hovels in Anacostia and payless furloughs, Waters, Howze, and their peers embraced their CCC duty and every training opportunity that came along. These challenges honed junior officer leadership skills.[32]

Leading the CCC organizations foreshadowed the inevitable problems of rapid mobilization that the army would later encounter. As a second lieutenant out of the 7th Cavalry at Fort Bliss, Howze served as a platoon leader in a CCC company that worked in the Apache National Forest, New Mexico. At first glance, CCC duty may seem dull to a company-grade officer, but the demands tested the men. The demands were so great that while Howze served as a CCC inspector in Georgia, he witnessed several officers and noncommissioned officers relieved of duty for unsatisfactory performance. The officers who led these CCC companies not only had to complete the projects, but also had to manage a very diverse labor pool in harsh conditions. Many officers viewed the work as arduous, but a welcome break in the routine of army training. Young officers thrust into command of groups of untrained and marginally disciplined men mirrored the experience of the early U.S. Army battalion leadership of World War II.[33]

The U.S. Army as a whole reaped numerous benefits from its participation in the CCC beyond the obvious military exposure introduced to millions of future World War II recruits and the positive public image of the U.S. Army gained through the cooperation with the public and civilian government agencies. The CCC effort required enormous manpower resources to manage 500,000 men spread out over 2,109 camps. To supervise this mass of humanity, Congress maintained the current active duty officer strength in a dire fiscal atmosphere. Some officers estimated that some congressional members desired to slash the officer corps by as much as 4,000. In the first two years, the army assigned approximately a fourth of its officers to lead the CCC companies. Eventually, this manpower requirement slowly whittled down to about 5 percent of the officers corps until 1940. In addition to the active army, up to 30,000 reserve officers—many out of work—found employment and experience leading the CCC around the country. Many future reserve officers would serve as battalion leaders in World War II; leading the CCC gave them their only leadership challenges beyond their two weeks of annual training. While the army did suffer recruiting hardships since CCC men received about twice the pay of most enlisted soldiers and CCC duty interrupted regular training exercises, the net benefit improved the regular and reserve army officer corps. The practical application of leadership encouraged initiative, innovation, and the mastery of logistics procedures. The company-grade officers of the early to mid-1930s often had the opportunities to ply their trade in the CCC as well as to develop their combat skills in annual maneuvers and branch-specific schools. Unfortunately, reserve officers often got to do only the CCC duty, but missed out on the more well-rounded careers enjoyed by active-duty officers.[34]

The U.S. Army assumed responsibility for almost every facet of the CCC's existence except the conduct of the actual work. Two to three officers ensured their charges received clothing, housing, transportation, entertainment, structure, discipline, finances, medical care, spiritual guidance, and education. Since traditional military law did not apply to CCC men, camp officers had to develop a broad repertoire of techniques to administer the camps. Seventy-five percent of the corps remained under the age of twenty-one. The relatively young nature of the corps exacerbated the enforcement of discipline, but presented a terrific leadership laboratory for army officers. The camp at Natural Bridge, Virginia, serves as an example. After five months with a nurturing leader, the CCC boys fell under the control of a young engineer lieutenant who led with more of an iron hand. The environment became so toxic that the lieutenant camp commander had to remain armed at all times. Furthermore, he had to contend with a severe drinking problem in the camp that culminated with a bloodless pistol duel between two drunks in front of the camp's gate. With little other choice, the lieutenant had to forcibly remove the troublemakers and send them home. For many officers, leading the CCC camps proved invaluable as their first command experience and taught them many lessons.[35]

Critics and supporters of the army's involvement acknowledged the U.S. Army's satisfactory ability to instill discipline and provide leadership. If they did not already face a daunting enough task, the company-grade officers had to coordinate this manpower to meet the project objectives executed by the politically appointed superintendents and the technical service foreman from organizations such as state park boards, road commissions, and the National Park Service. Camp leadership provided a structured environment that included an 0600 wake-up and reveille, a calisthenics routine, breakfast, a workday from 0800 to 1600 (with an hour break for a packed lunch), showers, retreat and dinner from 1700 to 1830, followed by educational and vocational studies, recreation activities such as sports, and free time, with lights out by 2200. As the CCC completed projects or received new assignments, the army had to construct new camps. Anyone who has visited a national park has seen the results of the CCC's efforts, but their work ranged far and wide to include soil conservation, road and bridge construction, biological surveys, digging of ditches and sewers, forestry, and mapping, to name a few. The dizzying array of projects and camp-management challenges forged key leadership experiences for young army officers stationed in places such as eastern Texas, southern Utah, or northern Wisconsin.[36]

Second Lieutenant Douglas Johnson's herculean exertions in the Wisconsin woods illustrate the CCC effect. Torn from his choice assignment working the World's Fair in Chicago, he was assigned by his adjutant to command the 657th CCC Company. Johnson mustered his advance element of three regular army noncommissioned officers and ten CCC men. In less than two days, the small band drew four vehicles, tents, rations, and anything they thought they might need in the camp. Next, they departed Fort Sheridan headed for Summit Lake, Wisconsin, without any money; they only knew that Summit Lake was in Wisconsin. After borrowing maps at a gas station and an illegal encampment at a baseball field after three sleepless nights, the advance guard of the 657th reached Summit Lake. They squatted at the local train station for the night, and Johnson linked up with the ranger of the state park. The work had just begun.[37]

Johnson had to establish a camp, provide essential services, and gain positive control of his company that still had not yet arrived. The ranger assigned a heavily wooded campsite located near a very poor road. To compound the problem, the camp had no water and no communications. Johnson had to travel 15 miles to the nearest telegraph and send his men to local farmers to ask for water. The next day, the wrong CCC company arrived at the railhead. After some commonsense haggling via telegraph, Johnson convinced his superiors to assign this company to him since it did not seem to matter much which group of men he led. Luckily, the newly branded men of the 657th were mostly teenagers and not the troublesome veterans who plagued the Natural Bridge camp in Virginia. Johnson saw himself as the "law and order" of the camp and committed himself to the idea that "nobody was going to take over that camp from me." Without the authority of the *Army Manual for Courts-Martial*, Johnson and his NCOs provided the structure required to turn these men into a viable workforce.[38]

In a six-month time period, Johnson used adaptive leadership to evolve his camp into a productive organization. Persuasive negotiations with local store owners provided food and equipment on credit until the corps wired him $1,500. The challenges never seemed to end. Johnson located a well driller to create a camp water source and an oil man to construct a fuel facility. He rebuilt the camp after a fierce windstorm. He even had to be a doctor. After one of the men sustained a grievous axe wound on his hand, Johnson located a doctor 15 miles away. The doctor could not complete the surgery, so Johnson sewed the man's tendons together and thereby closed the laceration. A month later, the man was catching on the camp's baseball team. Besides constant problem solving, Johnson still had to manage routine camp business and discipline. He developed an intramural baseball and boxing league, built volleyball and baseball fields, constructed hot water showers, and even organized a band and theater troupe. He maintained discipline with formations, barracks inspections, and routine hygiene. As a result, the camp had no sanitation issues and very few disciplinary problems. After a hard summer of Wisconsin State Forest projects and firefighting during a dry fall, the 657th got orders to move camp. Johnson moved his men and set up a new camp ready for the onslaught of a Wisconsin winter. A week before Christmas, Johnson left the CCC a better officer and leader: "The greatest benefit of all was, I believe, to the U.S. Regular Army. Its personnel got chances to exercise command in far greater numbers than would be normal in a peace time army, to operation supply activities of extended size, and to meet challenges of unusual situations. I know that nothing much ever seemed too big for me to handle after I returned from that C.C.C. detail, not only wiser, but wealthier."[39] Company-grade officers could not have asked for a more challenging leadership experience than the initial creation of the Civilian Conservation Corps companies. These junior leaders continued to build on their CCC leadership as they tackled annual maneuvers.[40] The CCC was only one crucible that impacted both the regulars and future citizen soldiers.

The New Deal instituted significant changes in American society that had lasting impacts on the wartime army. Various programs—including the CCC—employed almost 18 million Americans. The Federal Theater Project exposed over 30 million Americans to its productions. The average American citizen developed an internationalist and cosmopolitan outlook based on these New Deal initiatives. Essentially, citizens had no opinion on international affairs or even other regions in America before these expansive programs. The American people came to view the government as a competent problem solver that could mobilize massive programs and resources (9 percent of GDP in 1936). Federally sponsored programs such as the Works Progress Association (WPA) shaped this view, especially in a decade in which the unemployment rate stayed above 14 percent and the marriage rate dropped 25 percent.[41]

Without the New Deal, regular army officers would have struggled to develop combat-effective units and learning organizations as quickly as they did in North Africa and Sicily. The New Deal foundation began the process of producing a citizenry willing to accept the leadership of regular army officers. The vast mobilization of military forces continued to change attitudes. Soldiers typically traveled to four or five different training centers spread across the United States before deploying overseas. This travel further developed each soldier's sense of country and the tasks that lay ahead. Additionally, approximately a quarter of all veterans came into contact with New Deal programs, especially the National Youth Association and the Civilian Conservation Corps. The New Deal and mobilization experiences eliminated barriers between the citizen soldier and the professional regular army officers.[42]

Battalion leaders commanded organizations filled mostly with citizens drawn out of the New Deal society. These citizens included not only average soldiers but the junior officers who would provide direct combat leadership. The New Deal effects moved these new soldiers to accept the professional leadership they encountered in the U.S. Army. Army life prevented the New Deal—minus the CCC—from having a significant impact on the regular officer corps in the 1930s. However, the regular officers found common ground with their draftees and OCS graduates since both groups accepted American institutions as effective and legitimate organizations. The erosion of the barrier between citizen soldier and professional created units that could perform under the stresses of combat. When not in combat, these same units had cultures that promoted learning based on past experiences. Unlike the staffs at the division level and above stocked with significant numbers of regular army officers, battalions had to incorporate vast numbers of newcomers at every level but the field-grade positions. The New Deal in effect set the conditions for regular officers to lead their phalanxes of citizen soldiers.

The Thomason Act Diversifies

The 74th Congress recognized that the U.S. Army suffered from an acute manpower shortage in the mid-1930s, given the Civilian Conservation Corps duties. The dire state of the U.S. Army Air Corps also convinced the legislators to authorize an increase in officer numbers and training. What became known as the Thomason Act became law in the spring of 1936. This amendment to the National Defense Act of 1920 added 381 officers for service in the Army Air Corps due to the ever-increasing manpower requirements for new pilots. More importantly, the act—beginning on 1 July 1936—authorized the War Department to select 1,000 reserve second lieutenants for service with the regular army for a one-year term. From this pool, the army could select fifty officers to receive regular commissions based on competitive examinations and recommendations.[43] Originally designed to train 2,000 reserve officers of all ranks, as well as the 7,000 annual ROTC graduates, to distribute relevant experience throughout the ranks the act's advocates settled on the 1,000 second lieutenants per annum with the option to commission 50 as regular army lieutenants. At an annual cost of $1.85 million dollars in a time of squeaky budgets, the U.S. Army scored a major manpower victory.[44]

The Thomason Act created a symbiotic relationship between the reserves and the regulars. Professors of military science and tactics of ROTC units could select qualified officers to receive on-the-job training. After a year of experience and a salary that usually eclipsed that of civilian occupations for twenty-two-year olds, the officer returned to civilian life trained and richer. Many of these reserve officers also returned to active duty to serve the CCC. The regular army got to pick the cream of this crop of butter bars to augment the 275 graduates of the United States Military Academy.[45] Pure numbers and experience do not tell the full benefit of this program. The injection of greenhorn lieutenants required constant mentorship from the regular army officers as the act clearly stated:

> The careful supervision of the instruction and training of the reserve officers during their one-year of active duty is deemed a

> matter of the utmost importance. The general principle to govern this training is that whereas these officers shall enjoy the same status as newly commissioned officers of the regular Army in order to develop in them qualities of leadership and a sense of responsibility, their professional education will be systematically organized and efficiently conducted under the direct supervision of the commanding officer.[46]

The Thomason Act officers did not relieve the long-term commissioned officer shortage, but their injection demanded senior officers and peers to train their subalterns and interact with them.[47] At the company and troop level, their presence benefited the army.

Officers on both sides of the program praised its achievements. The program drew high-quality reserve officers into the regular army. Samuel A. Montague even left his job as Henry Ford's personal photographer to apply for the yearlong stint on active duty. After the army selected him, he applied and won one of the coveted fifty active-duty spots. From 1938 until World War II's culmination, Montague served in the infantry, the Army Air Corps, and the Combat Observer Corps, and guarded the Panama Canal Zone. Most reserve men believed they received professional treatment and development from the regulars, while the regulars regretted not having more of them. This constant exposure to fresh lieutenants developed key mentorship qualities throughout the regular army company-grade officers. As the army burgeoned with thousands of new officers in 1940 to 1942—both reserves and Officer Candidate School graduates—these future battalion leaders embraced the new officers. The erosion of a regular army bias prompted widespread acceptance of any officer. Regulars "realized that good officers can come from anywhere."[48] Their time with the Thomason officers prepared them to integrate and develop the novices who populated their battalions.[49]

The Next Step: Army Service Schools

Beginning in 1934, Major General Edward Croft, the new chief of infantry, ushered in a new intellectual focus for the Infantry School at Fort Benning. A former editor of the *Infantry School Mailing List*, Croft increased the emphasis on military history in the Benning curriculum and demanded a substantial increase in the quality of his house journal, *Infantry Journal*. The 1930s army has received much criticism for the parochialism of its branches mainly because branch chiefs such as Croft had almost unlimited power within their own branches.[50] Croft embraced this autonomy to increase the intellectual readiness of his branch. He also advocated significant hardware changes, including the most advanced individual infantry rifle of World War II, the M-1 Garand. Croft's initiative improved the quality of both the *Infantry Journal* and the *Infantry School Mailing List*. The *Mailing List* published material biannually from the Infantry School at Fort Benning to keep its branch informed on the latest thought and instructional material. The *Mailing List* had improved its format since 1930, but received even more relevant material once Croft weighed in at the school. The first issue of 1938 included substantial articles on motor reconnaissance patrols, motor patrolling, the utilization of aerial photographs for the combat officer, the importance and practice of motor maintenance, and the employment of mortars. This volume also included three historical vignettes on tanks at St. Mihiel, the crossing of the Tigris River in World War I, and Napoleonic discipline. A young combat arms officer could digest everything from the latest tactics and techniques to military history to the correct way to fill out an officer efficiency report.[51]

The *Mailing Lists* present an accurate representation of a young combat officer's professional military education. The Infantry School struggled to convince its officers to read its *Mailing Lists* while they conducted duties in their units. These combat officers eventually became familiar with these concepts in the *Mailing Lists*. Floyd Forte attended the Infantry School Regular Course in 1938 just prior to his doomed Philippine assignment, and Edwin Van Valkenburg Sutherland went in 1940, bringing his Philippine field experience to the course. Essentially branch-specific courses such as the Infantry School Regular Course served as an officer's first army school after graduation from West Point or ROTC. Branches designed the courses to instruct the officers in current doctrine, regulations, and new procedures. The primary instructors for the school not only taught younger officers like Forte, but developed their lesson plans into the *Mailing List* articles. Whether "Van" Sutherland read the *Mailing Lists* or not, he had to study the material during his tenure at the school. This intermediate education between West Point or ROTC and the Command and General Staff College functioned as a period of reflection about tactics for young officers.[52]

The January 1937 Infantry School *Mailing List* elucidates the curriculum that young infantry officers wrestled with at Fort Benning. The last three articles of this issue revolve around the light infantry defense against mechanized attacks. "Limitations of Tanks" begins the series with a whirlwind tour of the world's tanks. Clearly meant as an introduction to the various types of tanks, this article highlights the strengths and weaknesses of heavy, medium, light, Christie suspension, and amphibian tanks and half-tracks. "Antitank Defense," meant to be the centerpiece of the series, stresses the successful conditions for an infantry unit (regimental-size) to array its subordinates to defeat a mechanized assault. Relying on World War I French tank losses, which revealed an 88 percent destruction rate to cannon fire, the author asserts that antitank (AT) cannon must serve as the heart of any antitank-defense. The infantry unit must augment the AT cannons with artificial and natural obstacles. The unit's selection of the terrain dictates the logical employment of the guns and man-made obstacles such as tank ditches and minefields. Numerous data charts on armor thickness compared to cannon penetration and obstacle technique diagrams pepper the article. "Antitank Defense in Foreign Armies" lends a broader perspective to the two narrower studies. Relying on a vast array of German, French, and British field manuals, professional military journals, and World War I historic data, this final article lucidly compares and contrasts the three European powers' antitank approaches. Instructors at Fort Benning taught these concepts in their hot and stuffy classrooms at Infantry Hall. Thus, a company-grade officer in the infantry course received exposure to the latest technical capabilities of his own equipment and that of his foes. More importantly, these new ideas of mechanized defense must have seemed relevant to them especially when put in the context of foreign capabilities. Most annual maneuvers did not incorporate this latest tactical thought, but at least officers considered how they might conduct an antitank defense.[53]

The Cavalry School at Fort Riley lagged behind the more progressive thought of the infantrymen, but it had flashes of brilliance. Just as Croft had breathed new life into the Infantry School *Mailing Lists*, the Cavalry School took the opposite tack and ended its *Mailing List* at the end of 1933. Of the eight articles in the last issue of 1933, seven dealt directly with mechanization subjects. With appropriate subjects such as "The War of Tomorrow: The Motorization of the Cavalry," "Tactical Employment of Mechanized Cavalry," "Mechanized-Mindedness," and "Mechanized Forces (by Major George Patton)," the Cavalry School failed to exploit the use of its *Mailing List* as the Infantry School did.[54] Fortunately, the actual school nestled in the rolling hills of central Kansas implemented a rigorous nine-month course for its first lieutenants. Academic work dominated most of their time. Officers spent about four hours a day studying small-unit tactics and techniques as well as mechanization. All of them had to attend an automotive school and learn Morse code. Most of this technical training remained rudimentary, but it gradually improved throughout the 1930s. Unlike the whimsical duty days (when not conducting maneuvers) across the Pacific posts, students had no time for social engagements or sports during the training day. Students spent their weeknights working on tactical problems and reading just like at West Point, as some recalled. Even though most officers recognized that the horse had little use on future battlefields, the Cavalry School retained the horse as part of the curriculum for several reasons. Primarily, officers rode horses as a form of exercise. Since most posts retained substantial stables and officers learned to ride as cadets, the army viewed the horse as a natural outlet for physical training. At the Cavalry School, the students still spent about three hours riding their two assigned horses every day and studied hippology, veterinary medicine, and horseshoeing. As the 1930s progressed, the curriculum included fewer and fewer horse activities, but these never completely faded until 1941. The first lieutenants who cycled through Fort Riley endured a curriculum grounded in mechanization. Most of them came to view the horse as a recreational animal on which to play polo, compete for the Olympics and in horse shows, take country rides, and to fox hunt.[55] Despite the retention of the horse as part of the officer corps's culture, the cavalry thought about modern war.

The intellectual approach applied by the Cavalry School indoctrinated its officers with the demands of modern war. Throughout its curriculum, the school emphasized combined arms operations and movement in warfare. If anything, critics could level claims that the Cavalry School attempted to cover too much. The cavalry course leaned heavily on the campaigns of World War I to show the way to future warfare. Every conceivable cavalry mission received attention, from defensive combat to retrogrades to the attack to pursuit and exploitation. Rather than rest on the merits of historical vignettes, instructors hypothesized how the increased firepower, motorization, and mechanization of armies would affect the role of cavalry in the next war. The constant criticism of pre–World War I European cavalry's failure to prepare and adjust in the first years of the Great War served as a dire warning to young officers. Nearly a fourth of the case studies occurred during 1914, to underscore these failures. A cavalry officer understood his branch's role as a supporting element to infantry that brought "high mobility and fire power" to the battlefield.[56] The school probably emphasized movement and mobility too much at the expense of integrated execution with the infantry, artillery, aviation, and communications. Except for these major oversights, Cavalry School graduates grasped how the mechanization of their branch would require proper integration with all battlefield assets at a rapid pace.[57] Indeed, a lively debate existed in the branch between the eyes-and-ears traditionalists and the fighters. At the tactical level, both sides argued about the proper cavalry platform and missions of reconnaissance and security versus fighting economy-of-force missions. The downside of the debate was that it failed to develop a coherent operational cavalry doctrine. The two schools forced officers to wrestle with the doctrine at the tactical level.[58]

All of the service schools hammered home many of the same principles to teach officers how to train their soldiers and officers. To establish a common baseline comprehension, the Cavalry School's curriculum serves as an appropriate example, especially since it is often viewed as the most backward because of its attachment to the horse. The school incorporated a "training management" module throughout the nine-month course and even developed a textbook that students took away from the course. The training management course conveyed three crucial concepts: how to train, how to plan and evaluate, and types of training. The students encountered these concepts in their first assignment; however, they did not hold the responsibility for application as company officers. The "how to train" concept established the seven methods for training: map problems, map maneuvers, tactical rides/walks, terrain exercises, command-post exercises (CPXs), field exercises, and field maneuvers. Map problems (occurred in the classroom) and terrain exercises (occurred in the field) were designed to test officers' ability to solve notional tactical situations with a written solution and/or operations order. Map maneuvers and tactical rides/walks assigned both enemy and friendly roles in a tactical situation, and the participants discussed their notional actions. These first four events were not resource intensive and developed leader proficiency. CPXs, field exercises, and field maneuvers all required a more serious commitment from the units. The CPXs exercised the chain of command and staff to orchestrate a battle based on an order issued by the higher headquarters and fictional intelligence updates.[59] This event was critical to teach commanders and staffs to visualize the battlefield since they could never hope to see all of their soldiers. Field exercises and maneuvers both involved the employment of a unit in a tactical situation. The only difference was that the maneuvers included an enemy force and, usually, umpires to adjudicate the action.[60] "How to plan and evaluate" simply instructed students how to design regimental, squadron/battalion, and company/troop/battery training plans. The training schedule required officers to synchronize all the training with the appropriate exercises and higher headquarters events. Instructors stressed that officers continually evaluate their soldiers to determine when a unit gained proficiency in a task and when to dedicate more training time. After every exercise, no matter how trivial, officers should conduct a review of the event. Finally, the manual classified two "types of training" for soldiers and officers: mechanical or skill-based versus combat or application-based. Officers had to balance their training time between skill competence and combat exercises.[61] Not only had officers lived this training management in their first assignments, but after the service school, they could now execute it up to the regimental level.

The branch schools, regardless of their warts, trained company-grade officers in the latest tactics, military history, and leadership issues. When they left Forts Benning and Riley, these men were prepared to assume command of companies and battalions if necessary. All the schools inculcated the knowledge that leaders must lead by example, especially in combat. They all became experts on the tools of their trade, from engines to horses to machine guns. More importantly, they learned how to train subordinates in weapons and tactics. The intellectual development had the most lasting effect. Since the West Point curriculum was grounded in the tactical level of war, cadets did not read Clausewitz or digest the strategic level of war. The service schools elevated their thought above the tactical horizon and served as another conduit to encourage individual study of war. Some but not all officers read Fuller, Liddell Hart, and Clausewitz. At this early stage in their career, they wrestled with these concepts. This strategic exploration combined with their tactical expertise and operational study garnered from West Point. The company-grade officers' intellectual development prepared them to comprehend the linkages between the tactical and operational levels of war. They did not develop into strategic thinkers per se, but they could envision the complexities of operational design.[62]

Many officers, especially midcareer officers—the future general officers of World War II—viewed the branch schools, also known as service schools, as "the most important of all the schools in the military system of education."[63] They believed that neither the service schools nor West Point had prepared the lieutenants of the 1920s to conduct combined arms operations. Fortunately, both the service schools and

West Point had already heeded their advice just as they were speaking out. The service schools expanded the class sizes, and West Point evolved its curriculum. As the service schools expanded by the early 1930s, they could rotate officers through the service schools within three to six years after commissioning instead of the average of nine years. By the early 1930s, a lieutenant could expect to serve in one or maybe two duty stations before service-school attendance. The service school gave the officer branch expertise, prepared him to conduct "combined action" with other branches, and served as leadership development before the officer assumed company command. Some officers advocated another "basic" service school for new lieutenants or an additional year to the West Point curriculum.[64] This additional year would instruct new officers in their assigned branch. Few senior officers, however, seemed willing to invest another year into wet-behind-the-ears lieutenants.[65] Midcareer officers also encouraged service schools and West Point to increase mechanization and communications instruction as well as to stimulate individual study in subjects "related to the military art." Since the army did not create another level of schooling, the USMA filled the void. The academy's reformed curriculum did not make the cadets branch experts. It did provide a sound grounding in combined arms operations. Furthermore, the interdisciplinary-subject approach fostered an environment centered on the study of the military art. Whether in French class or a military engineering class, cadets were exposed constantly to the study of their profession. Hopefully, this early and frequent exposure to the military art inspired cadets to self-educate throughout their career. As one midcareer officer put it, "the time to begin individual study is when one enters the service."[66] If the academy accomplished anything, it certainly encouraged individual study of the military art. So, just as the future generals called for professional military education reform, the system had recognized the shortcomings and adjusted.

The Army General Headquarters Maneuvers

As early as September 1939, the U.S. Army's protective mobilization efforts began to set the stage for the comprehensive Army General Headquarters (GHQ) maneuvers of 1941, more commonly known as the Louisiana and Carolina maneuvers. On President Franklin Roosevelt's initiative to strengthen national defense, army chief of staff General George Marshall restructured the World War I–style square infantry division to adopt the triangular division. To complement this reorganization, the divisions were partially motorized and included artillery support for each infantry regiment. By the spring of 1940, these initiatives prompted the regular army to conduct corps-level maneuvers to test the changes. The Third Army field exercises at Fort Benning, Georgia, and Louisiana were the first corps-on-corps maneuvers in the U.S. Army's history. In addition to many lessons learned, the maneuvers spawned the armored force in July 1940. Prior to the 1941 GHQ maneuvers, the regular army had exposed its junior leaders to the challenges of combined arms warfare, although the army remained more concerned with division operations and above.[67]

These large-scale exercises, known simply as the Third Army maneuvers, occurred from 5 to 25 May 1940 in the Sabine Area on the Texas-Louisiana border; it was explicitly stated in the operations orders that "the combat training of small units is not their main purpose."[68] Generally, these maneuvers aimed to test the employment of corps of triangular divisions over vast areas against a mobile enemy under combat conditions that included combat aviation. Although training was not the stated purpose, commanders could not help but be trained. Unless they participated in Hawaiian or Philippine maneuvers, company-grade officers rarely saw anything like these movements. Not only did they witness them, but they had to keep their vehicles maintained, move in choking clouds of dust, traverse numerous watercourses, preserve unit discipline in all matters from field sanitation to march discipline, and, most importantly, translate guidance from their division commanders into actionable orders.[69] The large-unit focus of the Louisiana maneuvers obfuscated the value to company-grade officers.[70]

The maneuvers introduced junior leaders to a continuous stream of tactical problems. Infantry officers kept their soldiers loaded on trucks throughout artillery barrages and air attacks. An air attack caught a 155mm gun regiment on a road with almost no interval between vehicles. When the commander directed that the unit back up, chaos ensued, resulting in a monstrous traffic jam. Ground units could not synchronize their efforts with their air support. Ground personnel could not direct aircraft onto enemy targets. The vast distances strained maintenance readiness. Officers deferred maintenance in the field until the end of the maneuvers. Not surprisingly, operational readiness rates—the percentage of functional vehicles—declined throughout the maneuvers. Tank units often conducted attacks unsupported by infantry, artillery, air, or engineers. Furthermore, armored units failed to conduct proper reconnaissance and charged blindly into enemy defenses. Many observers concluded that "the training of small units and the individual soldier must receive more attention right away."[71] On the surface, these failures revealed glaring readiness problems at the tactical level. However, the junior officers who executed these poor decisions never forgot them. Throughout the maneuvers, commanders emphasized that all units execute flanking actions, up to 100 miles, over frontal actions. Furthermore, they demanded that subordinates attack at night and train combined teams of all branches and types of units.[72] Mobility dominated everything, so units maximized truck assets to move fighting men to the sound of the guns. For the first time, battalion leaders realized that it took about four hours for a corps order to reach a regimental headquarters and that a division required at least three and a half hours to clear a checkpoint or road intersection.[73] More often than not, units failed to meet the training standards and suffered numerous setbacks. The failures made an indelible impression on them not to repeat the mistakes.[74]

In the wake of the Third Army maneuvers and the stunning German offensive in France, the U.S. Congress authorized the twelve-month federal activation of 300,000 national guardsmen and reservists as well as 900,000 draftees for the army under the Selective Service Act. Brigadier General Lesley J. McNair, the newly anointed GHQ chief of staff, recognized the opportunity given to him to prepare this temporary army for a future war. He spearheaded an echeloned training system designed to build a soldier and his unit. The first part—the mobilization training program—dedicated thirteen weeks to individual and small-unit training. Combat training, the second training phase, developed regimental combat teams that were the building blocks of the triangular division over another thirteen-week to sixteen-week period. The third phase focused on division-level operations over a four-week period. The completion of the third phase paved the way for the capstone GHQ maneuvers in the summer of 1941. The maneuvers pitted the four field armies against one another in two Louisiana and Carolina phases and one Tennessee phase. Designed to replicate combat, the exercises were intense and incorporated as many combat functions as possible. Furthermore, the war games took place over varied terrain, from dry, rolling hills to bayous. A voluminous umpire guide and a small army of umpires adjudicated the faux combat.[75] The GHQ maneuvers are mostly credited with the development of the future World War II general officers that served as division commanders and above. Furthermore, the maneuvers exposed serious command and control deficiencies as well as shortcomings in combined arms integration, confirmed a flawed antitank and close-air-support doctrine, and revealed a severe training deficit for soldiers and junior officers. For the future battalion leader, however, the three-phase training program and GHQ maneuvers provided the ultimate environment for free leadership development and a chance to implement emerging doctrine.[76]

The second training phase prior to the GHQ maneuvers exposed the battalion leaders to critical training challenges of new recruits and officers. This phase required company and battalion leaders to transform phase-one soldiers into part of a functional combat team from the platoon to regimental level. Company, battalion, and regimental commanders were required to test individual soldiers or units on current training topics every day. The officers' duties eclipsed just these basic inspections. Since the thirteen-week individual training did not completely prepare each individual, officers incorporated "basic and technical training" concurrently with unit tactical training.[77] To maximize field time, officers implemented basic and technical training on every conceivable task for their soldiers. These tasks ranged from radio code practice to demolition to range estimation to vehicle identification. Beyond the inspections and augmented individual training, the officers tackled the heart of phase-two training: the tactical field situations. Officers were required to conduct a two-part approach to the tactical situation. In the first stage, officers created a tactical problem such as the defense of a hill and fought the problem on an aerial photograph or map. After they established the training requirements and various solutions, the command team executed the same scenario on a sandtable. Based on the scenario, the unit then conducted the same exercise on the terrain depicted on the photo/map and the sandtable. Once the unit officers believed that their units were familiar enough with the tactics, the unit executed the second stage, with similar missions on different terrain under different conditions. The units were not permitted to rehearse these second-stage missions. However, officers could repeat the missions to emphasize certain learning points. Units added to the training's complexity by requiring that each unit conduct at least one night operation per week. In addition, the training guidance mandated air-ground training, realistic supply procedures, forward observer training for all maneuver officers, and platoon leaders of different branches to switch leadership positions during exercises.[78] The second-phase training demanded that officers plan and conduct training. After the intensive unit combat training, the officers moved into phase three—division training—and then on to the ultimate test, the GHQ maneuvers.

If any company-grade officer remained unconvinced how warfare had changed during his career, the Louisiana, Carolina, and Tennessee maneuvers put those doubts to rest. Several officers remembered this training as the most "demanding" of their careers.[79] Units moved hundreds of miles in less than two days. An observer of the second phase of the Louisiana maneuvers traveled just over 1,100 miles in four days and wore out four different jeep drivers from the 3rd Armored Division.[80] Leaders had to maneuver their units at night in all sorts of vehicles from trucks to tanks. The pace tried the young leaders. For weeks, officers not only had to move at night, but had to fight all day long. Unlike the more simplistic maneuvers of their earlier careers, usually set to a rigid timetable, the GHQ army maneuvers showcased what combined operations would demand of battalion leaders. These GHQ maneuvers were similar to the regular army Third Army maneuvers, and battalion leaders gained familiarity with the same lessons learned. Some issues, such as vehicle maintenance, showed vast improvement from the previous spring. Problems still persisted though. The overwhelming criticisms at the tactical level centered on poor unit reconnaissance and cavalier concealment from enemy aircraft.[81] Indeed, General McNair listed "small unit training and inadequate officer leadership" as the principal weaknesses revealed by the Louisiana maneuvers.[82] McNair advocated the removal of weak officers—at all levels—who could not command or instill discipline.[83] "Where a weak leader can be replaced by a strong one, the necessary action certainly should be forthcoming without hesitation on the part of higher commanders, as soon as the situation is clear. And time is fleeting."[84] While many critics point to these glaring errors as indicative of future combat performance, they miss the entire point of training. Battalion leaders realized these shortcomings during the GHQ maneuvers. They struggled to overcome them. The maneuvers were viewed as the most realistic test in preparation of war, and commanders did not tolerate feeble leaders.

These junior leaders lived with many of the shortcomings identified throughout the maneuvers, but they could address only those within their power. Commanders and observers concluded that tank and antitank battalions must have direct communications with supporting aircraft.[85] A battalion-level leader could never hope to fix this systemic problem between an almost service-separate air force bent on strategic airpower and the U.S. Army. The army and the air corps never truly solved this problem until Operation COBRA and the subsequent breakout in France in the summer of 1944.[86] Logistically, the maneuvers were usually well supplied. The movement of supplies, however, relied heav-

1940 Third Army Maneuvers, East Texas. This is a photo taken in the vicinity of Jasper, Texas, during the 1940 Third Army Maneuvers. The low-angle view depicts the broken woodlands of east Texas. 4th Corps Provisional Observation Group, 1940, "Photos of Training Areas in Texas, Louisiana, and Georgia," 187 photos and 23 Maps, CARL.

ily on railroads and multiple road networks. The anemic rail and road network in North Africa bared this operational tether. Like the close-air-support issue, this logistical Achilles heel lay beyond the battalion leaders' radar. On the contrary, a battalion S-3 (operations officer) who witnessed nine of his tanks picked off by an antitank unit could learn to coordinate his attack in the future with artillery and infantry.[87] The larger institutional challenges took much longer, admittedly, to adjust than the tactical lessons learned. Tactical leaders could immediately adjust their training of a particular exercise. They could also evolve their solutions to tactical problems during and immediately after the exercises.

Donald Bennett, USMA class of 1939, served as a battalion S-3 during the Louisiana maneuvers with minimal experience. With only two months as a battery officer and seven weeks of basic artillery training, he found himself embroiled in the largest maneuvers to date. One incident during these maneuvers articulates his leadership development.

Bennett's battalion had stopped to provide fire in support of a bridge crossing. While milling about the battalion operations center, Bennett noticed the battalion sergeant major standing in the middle of the command post with a "1000-mile stare." The sergeant major did not even look at Bennett, but spoke as if to anybody listening: "The battalion commander has just been captured." Bennett, already aware of the capture, simply nodded. The sergeant major then uttered, "The battalion executive officer is lost." Again, Bennett knowingly nodded. Finally, the senior NCO said, "Somebody has to tell this battalion what to do," and spun around and walked out of the command post. Bennett assumed the leadership reins and completed the mission as the commander. He remembered that he "graduated that day" and assumed that the sergeant major had enough faith in him to lead the battalion or he never would have left the tent.[88] Bennett's temporary tenure as the battalion commander offered him a "free" leadership lesson. Although he was short on age and experience as a 1939 graduate, the scale of the maneuvers and his responsibility shaped his leadership.

Commanders at the division level and above recognized that they needed to disseminate information rapidly to their mechanized units. General McNair said, "Field orders in many units still were too long and involved, and yet lacked clarity."[89] He emphasized that units must rely on operational maps and overlays, standing operating procedures, short orders, and oral orders when applicable.[90] During the maneuvers, staffs at all levels gradually developed an abbreviated technique to command and control units. Exposure to this new technique allowed officers to execute missions based on very little guidance. This new style was a clean break with World War I standing operating procedures. A typical American Expeditionary Forces (AEF) division attack order in France in 1918 was fifteen to twenty pages long. Even after significant efforts, battalion orders averaged four pages in length as late as 1927. Eventually, U.S. Army officers gradually reduced operations orders to one to two pages and even issued them only orally. Communication with German and Soviet officers and consumption of their professional journal articles slowly convinced U.S. officers to change their ways. The Louisiana and Carolina maneuvers provided the vehicle to fully implement this new operational technique. Throughout the maneuvers, divisions produced simple operational statements to fight the simulated battles. This technique forced officers at all levels to inculcate unit SOPs and operate with very little guidance. Divisions continued this trend when they penetrated the Axis defenses in the Mediterranean.[91]

The maneuvers and battlefield developments in Europe prompted a significant review of army doctrine. By the mid-1930s, both officers and service schools had identified the problem that the U.S. Army did not possess a clear capstone manual for operations. The mounting pressure forced the development of a tentative *FM 100-5 Operations* manual in 1939. Drawing on established curriculums, both the Infantry and Cavalry Schools contributed heavily to the development of the draft manual. The manual emphasized maneuver and offense, but it still was infantry-centric. However, the Louisiana and Carolina maneuvers and German victories in Poland and France forced significant revisions of this tentative manual before dissemination to the field. The new version, *Field Service Regulations FM 100-5* (1941), directed that the U.S. Army employ the concept of combined arms warfare. This concept primarily advocated offensive operations and two forms of attacks—envelopments and penetrations. Commanders employed three typical forces to execute attacks: a main attack, a secondary attack, and a limited reserve. These combined arms attacks were to occur along a narrow front. To maximize room to maneuver and firepower, a unit synchronized all of its assets: infantry, tanks, artillery, engineers, tank destroyers, and combat aviation. Successful attacks facilitated access to enemy rear areas as well as pursuits to destroy enemy forces.[92] The branch schools' curriculum had advocated similar doctrine to the 1941 *FM 100-5* even though the U.S. Army did not codify it until a year before units landed in North Africa.

The 1941 *FM 100-5* contained infeasible doctrine and had some serious omissions. For example, the manual did not provide any guidance for amphibious landings or assaults. Battalion leaders filled the doctrine gap with practical experience, rehearsals, and extrapolation of other relevant doctrine, such as chapter 12, which referenced attacks on fortified positions.[93] Its discussion of tank destroyer doctrine became impossible to employ—with very few battlefield exceptions—given the nature of combat and the limitations of tank destroyers. According to doctrine, tank destroyers would mass firepower to defeat enemy penetrations. Their deficient armor and armament coupled with actual combat conditions prevented commanders from using them as directed in *FM 100-5*. Commanders usually assigned tank destroyer units as part of battalion-sized task forces to protect flanks. In both instances, battalion leaders improvised. The battalion leaders lived through a tumultuous time of ever-evolving doctrine. Hence, doctrine served as a starting point to visualize the battlefield. Combat operations dictated innovative application, and leaders could not wait for formal dissemination of doctrine updates.

As doctrine changed so did the organization of the army's infantry and armored divisions. Generally, infantry divisions performed all types of operations and received the requisite reinforcement when the mission required additional forces such as tanks. Armored divisions offered mobility and shock action to execute breakthroughs, exploitations, and pursuit of enemy forces. Both types of divisions had three regiments, giving rise to the so-called triangular division. The triangular division organization provided three regimental headquarters to coordinate administrative, training, and maintenance issues and three combat commands (CCs)/regimental combat teams (RCTs) to create combined arms task forces organized for combat operations. For example, a division could normally create nine battalion-sized task forces utilizing each battalion (nine in a division) headquarters as the command and control element. Depending on the mission, the division would assign these battalion task forces under the three combat commands. Infantry and armored divisions used these three combat commands to execute every type of operation outlined in *FM 100-5*. While this organization gave battalion leaders considerable flexibility, the divisions had shortcomings. Both infantry and armored divisions included underarmored reconnaissance battalions designed for reconnaissance and surveillance. They could not fight for information without augmentation. Additionally, both divisions had poor antitank guns that further undermined tank destroyer doctrine. The infantry division did not have adequate trucks to move its soldiers and lacked organic tank support. These conditions created unnecessary confusion and forced infantry divisions to coordinate at the corps level to move mounted or incorporate tank units. Moreover, the lack of established infantry and tank unit relationships often led to suboptimal combined arms teams on the battlefield.[94] Ultimately, the triangular structure tailored units for the mission, and battalion leaders became comfortable with mercurial organizations. An officer's grasp of doctrine and unit organization coupled with maneuver experience all contributed to the battalion leaders' bedrock knowledge of their profession. An effectual professionalism ensured that these leaders lived up to a rigorous standard. Their efficiency reports tell the tale.

The U.S. Army required annual efficiency reports for every officer to assess his performance in relation to his peers and potential for future service. All professional organizations strive to insulate themselves against report inflation. The U.S. Army is no exception; however, the interwar army policed the company-grade ranks well. This professionalism fostered company-grade officer adherence to high standards. The army found it difficult to force substandard officers out of service unless they committed grievous capital crimes. However, substandard officers consistently received poor ratings. When looming war clouds spurned the army's 1940 mobilization, General Marshall seized the opportunity to retire approximately 500 incapable field-grade officers.[95] Their forced retirement was mostly based on two things: age and efficiency reports.[96] This interwar army officer deadwood is often cited to paint the entire officer corps as stagnant, unimaginative, and dogmatic.

The four iterations of U.S. Army efficiency reports since World War I have one bottom line rating to assess an officer. Forms often included narratives as well as boxes to highlight certain character traits, but any superior evaluating an officer will focus on the overall rating given the officer. For the interwar army, efficiency reports included five key ratings: superior, excellent, satisfactory, unsatisfactory, and inferior. Interwar War Department reports indicate a healthy rating system especially of company-grade officers.[97] A healthy rating system is defined as one that does not hyperinflate ratings so that primarily only one rating—usually the highest—is employed to rate the vast majority of the officer pool.[98] In 1936, 28.2 percent and 22.1 percent of infantry and cavalry officers, respectively, received superior ratings; 64.2 percent of the infantry branch and 73.8 percent of the cavalry branch were judged excellent on their reports; 7.3 percent of infantry officers earned satisfactory ratings, as well as 4.1 percent of cavalrymen. Finally, less than 1 percent received unsatisfactory or inferior rankings.[99] As a comparison to the two maneuver branches, the Adjutant General (AG) Corps rated 72 percent of its officers superior and 27 percent excellent, and the Judge Advocate General (JAG) Corps had 36 percent listed as superior and 62 percent as excellent.[100] Both the AG and the JAG Corps drifted outside of healthy evaluations compared to the infantry and cavalry. Regardless of branch, the lieutenant colonel and colonel ranks were bloated with far too many superior rankings. In order to determine the relative efficiency of officers, the War Department even developed methods to compare officers across branches, utilizing branch averages. Based on relative values established by computing cumulative annual averages, commanders could judge the "relative value" of an officer compared to those of his branch and other branches.[101] Determining the relative value of an officer is a murky proposition, but the War Department's efforts indicated a serious commitment to officer professionalism. More importantly, the rating trends confirm a healthy professional environment for those officers in the grade of captain and below.[102]

Company-grade officers seemed to have benefited from a more balanced assessment process aimed at honest evaluations. Throughout the 1930s, captains and lieutenants from the Philippines to Fort Benning operated in a profession regulated by a healthy rating system. Approximately 10–20 percent of company-grade officers earned superior ratings; 65 percent received excellent; 10–15 percent got very satisfactory; and about 5 percent or less wallowed in satisfactory or unsatisfactory ratings.[103] Upon first glance, critics might cry foul at the prodigious application of the "excellent" rating. In reality, the efficiency report boiled down to only four ratings since both the satisfactory and unsatisfactory ratings marked an officer with the same negative connotation. Senior officers valued the report as an important tool to delineate the exceptional (superior) officers from the mainstream (excellent) ones and the ones who required more mentorship or maturity (very satisfactory). A satisfactory/unsatisfactory classification clearly indicated that an officer should not be an officer. Almost 20 percent of infantry and cavalry second and first lieutenants ended up with very satisfactory ratings, indicating two trends. A young officer often struggled to adapt initially to military life, and a commander could employ a lower ranking to make the rated officer self-aware. The young lieutenant could then adjust and/or improve to earn a higher ranking the following year. On the other hand, the excruciatingly slow promotion rates kept substandard officers at company-grade ranks for well over a decade. Underperformers absorbed many of the very satisfactory ratings so they would not advance to lead larger organizations.[104] Division commanders could identify quality officers to serve as their battalion leadership based on these accurate, healthy evaluations. Certainly not all officers lived up to past ratings, but the data indicate a consistent system aimed at identifying professional leaders.

No two officers had the exact same career path after their graduation from West Point or from their ROTC program. The experiences described above must be viewed as the most common occurrences throughout an officer's prewar career. Certainly not every officer led a CCC unit à la Douglas Johnson in Wisconsin or moved 600 horses like Hamilton Howze did at Fort Oglethorpe. The 1930s did offer more leadership opportunities like the CCC to company-grade officers than their peers had a decade earlier. They were well grounded in how to lead units at the battalion level and below because they had done so or seen it done. Their practical knowledge complemented the reflection and skills gained at their service schools. By the time these officers led World War II battalions, they had participated in maneuvers at all lev-

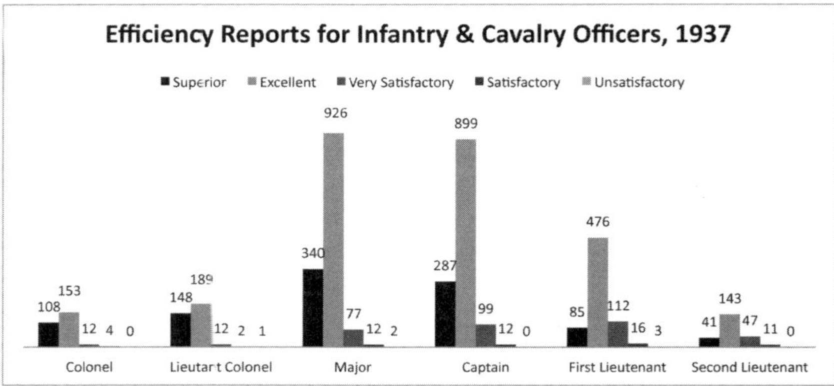

1937 Officer Efficiency Reports. Classification of Officers—Regular Army Based on General Efficiency Records, 15 March 1937, War Department, Statistics Branch, 1917–1945, Special Statistical Reports, No. 247, RG 165, NARA II, 1–9.

els several times over. They had executed orders with little to no oversight and minimal written guidance. Furthermore, the GHQ maneuvers terrain as well as that of Hawaii, Panama, and the Philippines required leaders to maneuver well beyond the line of the sight of subordinate and superior units. The maneuvers' speed and unit diversity introduced the officers to the rapid pace and complexity of combined arms warfare. Instead of just reading about German advances in Poland, France, and the Soviet Union, they saw it and acted it out for themselves. The maneuvers became one of the formidable learning events in their professional lives. They would not get many more chances before the penetration of the Mediterranean basin. During this mobilization and training, the Japanese attack on Pearl Harbor ensured U.S. entry into the war and removed any doubt that the U.S. Army would apply this training in combat. The "Germany First" pillar to Allied grand strategy determined that these battalion leaders would begin the war against the Germans and Italians in North Africa.

3

Wading in with TORCH

In order to comply with a Germany First strategy, the Allied effort required an invasion to confront the Axis armies. While the American Joint Chiefs of Staff argued for an invasion in 1943 of occupied France using the BOLERO/ROUNDUP plans, the British and President Franklin Roosevelt advocated landings on French North Africa. With American public opinion in mind, Roosevelt agreed with his British allies for political advantages in upcoming midterm elections, since chances of a slaughter of his untried countrymen seemed less likely against the French. This decision also assured the American public that the United States would emphasize the war against Germany first and funnel the majority of resources to the European theater. The intent of the operation centered on putting an overwhelming force off the Moroccan and Algerian coasts that would intimidate the French into waving a white flag. Once ashore, the units would combine to form the British First Army and advance eastward to confront Field Marshal Erwin Rommel with his Afrika Korps and Italians, thereby squeezing them against the British Eighth Army in the west.[1]

Three task forces formed the invasion force, and they dwarfed previous Allied amphibious operations at Dakar, Dieppe, Madagascar, and Guadalcanal. The landings would occur at three different sites with good ports and airfields: Casablanca, Oran, and Algiers. Despite British insistence that they could control the 8-mile-long Gibraltar Strait with little Axis interference, the Americans insisted on a landing site on the Atlantic coast in case Axis aircraft and submarines sealed off the strait. The Western Task Force consisted of American ships with American soldiers and sailed 2,000 miles from Virginia to the Moroccan coast. Command of the task force (to become Fifth Army Headquarters after landing) fell to General George S. Patton. His task force divided into three subordinate commands to control the widely separated landings

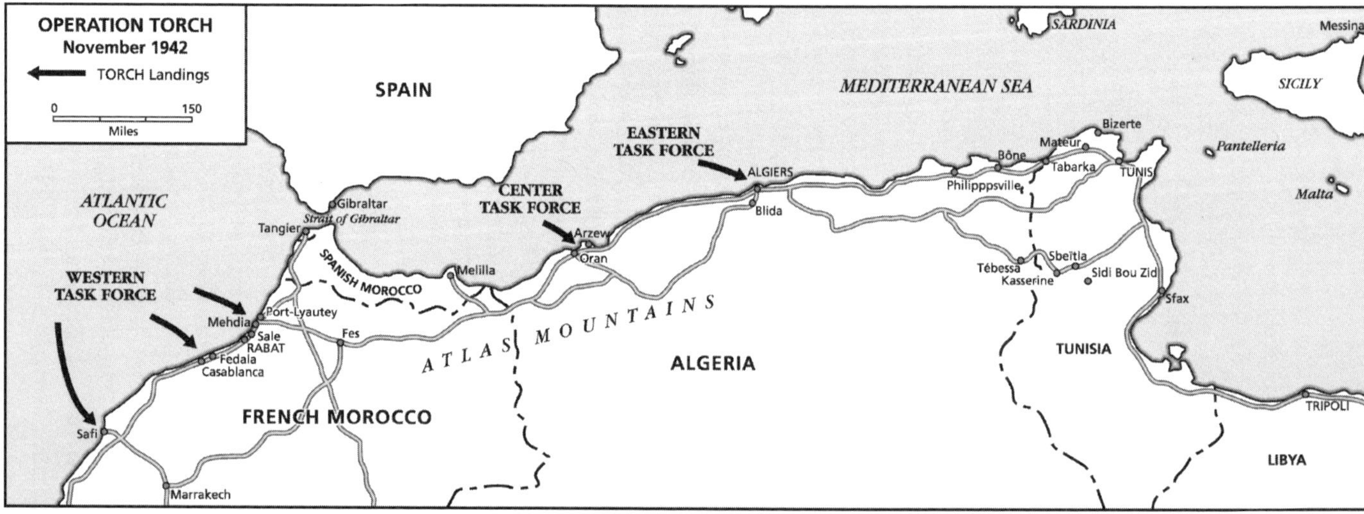

Operation TORCH Overview. From *Algeria–French Morocco*, The U.S. Army Campaigns of World War II, Center of Military History Publication 72-11, n.d., 6–7.

surrounding Casablanca.[2] The vast majority of U.S. troop strength sailed with this task force directly from the United States.

The Royal Navy would provide the majority of the ships for the Center Task Force setting sail from Britain, but carried American soldiers under the command of Major General Lloyd Fredendall. The 1st Infantry Division commanded by Major General Terry de la Mesa Allen, Task Forces Red and Green made up of components of the 1st Armored Division operationally controlled by Brigadier General Theodore Roosevelt, Jr., and the 1st Ranger Battalion commanded by Lieutenant Colonel William O. Darby made up the troop formation for the Oran, Algeria, landing force. The Eastern Task Force mimicked the Center Task Force in that most of the ships used for transport were British and the majority of the soldiers were American, under the command of British Lieutenant General K. A. N. Anderson. Both the Eastern and Center Task Forces employed amphibious assault against the main port facilities of Oran and Algiers and weighted their main efforts with landings on both flanks to envelop the ports, forts, and surrounding airfields.[3]

A superficial review of the Allied plan held promise for the planners and optimists that the target of French North Africa would not be a bloody adventure. All agreed that landing forces must be prepared to fight the French, but all the planners and commanders hoped that the French would capitulate quickly and peacefully, even though the French record prior to TORCH should have signaled violent opposition to the Allies. The French colonial army had fought the British in Senegal, Syria, and Madagascar prior to TORCH.[4] The Allies should have anticipated a fight no matter what the size and composition of their task forces, but the overriding issue concerned the disposition of Axis airpower and seapower in the Mediterranean and west of the Strait of Gibraltar.[5]

Neither the Germans nor the Italians predicted an Allied invasion of the western Mediterranean by mid-October 1942. Unable to neutralize British-controlled Malta, the Afrika Korps hung on by a tenuous logistical lifeline and remained relatively static while facing the British Eighth Army. The Allies achieved complete strategic surprise, and the Axis powers never detected the Western Task Force at all. U-514 spotted part of the Eastern Task Force west of the Strait of Gibraltar, but reported the incident as a routine supply convoy.[6] By the time reconnaissance planes discovered the armada in the western Mediterranean, Hitler made the decision to focus his thirty-five submarines and seventy-six aircraft in the Sicilian Strait to thwart the suspected landing sites of Tripoli or Benghazi, which he wrongly thought were designed to cut off the Afrika Korps from its supply lines. The Ultra evidence confirmed that the Axis powers did not suspect French North Africa as the potential landing sites, assuaging the fears of those concerned with lurking wolfpacks or concentrated German aircraft.[7] The ability to deceive the Axis powers and the dispersed landings prevented any real interdiction of the landing forces save for two ships damaged by lone Axis submarines. The TORCH landings caught Hitler and Mussolini flat-footed and denied them the opportunity to interrupt the landings with any significant forces.[8]

Operation TORCH and the subsequent North African campaign present the ideal situations in which to evaluate battalion leadership since they served as the first major combat tests for the U.S. Army in World War II.[9] During the landings and movement east to Tunisia, battalions often fought as separate entities or part of combat commands with little to no division oversight. Logistical shortcomings and convoluted command structures threw many units piecemeal into battlefield hot spots. TORCH also employed every battlefield organization and mission developed by the rapidly mobilizing army. From the Algerian/Moroccan beaches to Tunisia, regular and National Guard infantry divisions, two armored divisions, an airborne battalion, and a ranger battalion executed opposed and unopposed amphibious landings, offensives, and defenses against three different countries of varying capabilities, all in less than four months.

The Western Task Force Arrives in Morocco

The Western Task Force was laden with senior-officer talent who lived to develop effective organizations. Men such as Lucian Truscott, Ernest Harmon, and Maurice Rose led the units and staffs that sailed from the United States to the Moroccan coast. Their task force was an insurance policy to secure a lodgment west of Gibraltar. Senior leaders would have little to fear from the French. Instead, they encountered systemic problems that degraded combat capability. Of the three landing sites, the French offered significant resistance only at Mehdia and Port Lyautey north of Casablanca, largely because the attack began two hours after dawn. When not fighting the French, the Americans were battling failed systems and training or weather. Surf swelled up to 10 feet, wrecking landing craft, drowning soldiers, and sending equipment to the bottom of the sea.[10] Division commanders like Truscott lost any situational awareness since all communications from ship to shore failed save for a few naval gunfire shore parties.[11] Battalion-level officers and below never had a chance to conduct unit rehearsals and synchronize their staffs, nor did they get any attachments until boarding their ships.[12] Army and naval personnel all suffered from inadequate landing operations experience, which resulted in the random scattering of troops on the landing beaches.[13] Shore parties could not organize supplies or rapidly unload needed items. Moreover, medical support and evacuation simply did not occur above the regimental level.[14] The elements of the 3rd and 9th Infantry Divisions and the 2nd Armored Division that comprised the battalion landing teams (BLTs) had to overcome daunting shortcomings.

The Western Task Force's plan defaulted to the BLTs' initiative, execution of the plan, and rapid adjustment to frantic regimental and division commanders bounding around the battlefield issuing verbal orders. The landings at Fedala relied upon ten BLTs organized into three regimental landing teams (RLTs) under General Patton's command.[15] Artillery, snipers, and small French tank detachments offered the most trouble for the U.S. soldiers. In almost all instances, the BLTs destroyed French resistance with little assistance from their headquarters even against fortified artillery batteries such as the ones on Point Fedala.[16] The landings at Mehdia/Port Lyautey created the most chaos and challenges for the landing force. As the Americans landed late, scattered, and in daylight, the French reacted and were able to defend the fortress known as the Kasbah. Reducing fortified strong points such as the Kasbah presented the most significant problem for the northern landing force, mostly because the attackers lacked effective communications, artillery support, or any reliable air support.[17] Battalions relied on their own organic assets and personnel to fight the plan and orders transmitted, usually verbally, by their commanders.[18] Whatever the struggles or delays, battalion leadership displayed courage—sometimes blindly in frontal assaults—in leading their men with such uncertain support.[19]

Across the Algerian and Moroccan coasts, battalions executed their missions to secure French surrender by 11 November and were plagued by the same deficiencies in planning, training time, and coordination. Yet, battalions were led bravely, and eight valor awards were handed out to battalion commanders. Certainly, the French represented the "minor league" compared to the seasoned Germans lurking in the east, and Allied forces severely outnumbered the Vichy forces.[20] Battalion commanders, however, led their forces severely handicapped in regard to planning and training time. Furthermore, the Allies had stepped in almost every amphibious landing pitfall and still secured their objectives largely due to the ability of battalion leaders to execute commanders' intent. Combat observers determined that "the Infantry of the 3rd Division appeared to be determined, well-trained, and well-officered. The commanders [of battalion landing teams] seemed to know exactly where they were to go and in cases where previous plans went awry they used their initiative to straighten the situation."[21] Senior leaders such as Truscott served competently, but recognized the almost complete failure of their staffs to mitigate combat friction.

Fortunately, the average American GI showcased his usual propensity to solve tactical problems.[22] Although every combat observer and commander commented on poor soldier discipline and inadequate training throughout the landing, the soldiers' readiness was a product of minimal unit preparation as well as limited adjustment to the army. When it counted, the average enlisted man adjusted to combat. For example, U.S. soldiers displayed admirable aptitude in their rapid adoption of the bazooka. A few solders received very basic instruction on its use during the journey from Virginia to Morocco, but they did not get to actually fire the weapon. Troops in combat were eager to employ any weapon or technique that would increase their chance of mission success and survival. Several units used the bazooka in the attacks against the Kasbah fort in Mehdia and French tanks. In one instance, an army officer rounded up a dozen sailors on shore party duty, collected a few bazookas abandoned on the beach, and gave the hasty detachment less than five minutes' training. Shorty thereafter, the sailors-turned-soldiers scored a hit against a Renault tank.[23] While it was true that soldiers were mostly green and had devoted little thought to war prior to 1942, they served well enough throughout TORCH to support their battalion efforts.

Mixed Results: Elite Units at Oran

In September 1942, General Mark Clark briefed his airborne planner, Major William P. Yarborough, on Operation TORCH. In all of western Algeria, there were only two quality airfields, Tafaraoui (hard surfaced) and La Senia, located outside of Oran—the objective for the Center Task Force. A vertical envelopment of the two airfields seemed like the ideal mission for the budding airborne force since they could both neutralize the French air forces from interfering with the landings and preserve the airfields for later Allied use. Yarborough, who would serve as the executive officer, informed the 509th Airborne Battalion commander, Edson Raff, of the mission he called "the most important thing in your life," and both men went to work planning and training.[24] Raff embraced the mission completely and declared that the mission would succeed if the weather was good, if the air corps was effective, and if he was "permitted to command" his own paratroopers when they "hit the ground."[25] Both officers' attitude and excitement over the mission show young and motivated leaders eager to lead men in combat.

Raff developed a solid training program in conjunction with the British First Airborne Division while training in England and Northern Ireland. The 509th Battalion benefited from working with the British. The U.S. Army had a propensity to simulate combat conditions too much in training exercises. In one instance, a British umpire prevented the 509th from achieving their objective of blowing up a bridge since they did not bring along the appropriate demolition kit. The American soldiers were outraged, but quickly learned that saying "Well, we'll blow it up" was not going to work in training or combat.[26] The British passed on their proclivity for extremely precise terrain models to the American airborne as well.[27] The officers appreciated that high-quality terrain mock-ups of a target area were essential for memorizing objectives and surrounding terrain.[28] Unlike their brethren sailing on ships, Raff and his men conducted full rehearsals and numerous model exercises—with only the names removed—six weeks before they executed the operation.[29] With British help, Raff had utilized effective training management to transform his band of volunteers into a well-trained outfit. The 509th, nicknamed "Geronimo," were ready for the first U.S. combat airborne operation in history.

Despite the tactical readiness of Raff and his paratroopers, Operation VILLAIN failed to meet its intended objectives. By the U.S. Army Air Force's own admission, VILLAIN's collapse rested primarily with the air procedures—a problem that would continue to haunt soldiers jumping from the sky until the end of the war. Transport pilots were shockingly unprepared and had not even developed simple standing operating procedures for airborne operations such as nighttime navigational aids.[30] They had not rehearsed long-range instrument flights at night, and they did not have assigned navigators.[31] Lacking the homing beacon that was broadcast on the wrong frequency by a British ship and shut down too early by a U.S. State Department official in Oran, some planes landed in Spanish Morocco, others dropped their paratroopers near the objective, and still others had to land in the desert since they ran out of fuel.[32]

Before the 509th even arrived, amphibious forces seized the airfields. When Raff jumped, he suffered two broken ribs, which forced him to seamlessly pass command to Yarborough. Yarborough then took an ad hoc force to assist with French prisoners at Tafaraoui airfield. French Dewoitine D.520 fighters riddled Yarborough's three C-47 transports, forcing them to land back on the desert floor.[33] Seemingly, VILLAIN was a complete disaster. Both leaders, however, managed the unbelievable chaos and demonstrated significant maturity given the dire circumstances. After Yarborough's air-to-air contact, Raff—after an hour's rest with his chest wrapped by the battalion surgeon—led his unit to link up with Yarborough, and they all consolidated on Tafaraoui airfield.[34] There, they secured the airfield and guarded prisoners, which freed up 1st Armored Division forces to finish the capture of Oran. Vertical envelopment failed to take the Oran airfields, but the first U.S. commando unit—1st Ranger Battalion—fared much better in its assault against the Oran harbor gun batteries.

Attached to the 1st Infantry Division and supporting the seizure of Oran and the Z landing beach was the 1st Ranger Battalion led by Lieutenant Colonel William O. Darby. Prior to the 1st Infantry's landings, Darby and his rangers rapidly captured Fort de la Point, Fort du Nord, and Fort Superieur and rendered the French guns inoperable by 0330. Darby's executive officer, Major D. W. Dammer, conducted the first assault against Fort de la Point to distract Fort Superieur's occupants. Dammer's two companies' lightning-quick movement subdued the French in Fort de la Point while Darby hustled with his remaining four companies up the hills to Superieur. The element of surprise had evaporated with Dammer, so Darby emplaced his mortars and attacked with three companies abreast against the sunken gun emplacements. Withering French machine-gun fire met the advancing rangers, so the initial assault stalled. Darby recognized the precarious situation, ordered his men to temporarily retreat, and orchestrated a sixty-round preplanned fire mission against the enemy position. The machine gun ceased to bark. The Rangers completed the assault on Superieur and moved to Fort du Nord, which surrendered without a shot.[35] The rangers had completed their mission, but lacked the means to signal the assault force waiting 5 miles offshore. When Darby's landing craft was being lowered prior to the assault, the brake on one side stuck and spilled the men and their equipment into the water. Darby lost all of his white flares and the long-range radio needed to signal the fleet. As a result the fleet did not launch the assault force until two hours after the rangers' actions.[36]

General Marshall had ordered the formation of an American commando unit in late May 1942. Darby, only thirty-one and still a major, got the nod to assemble the 1st Ranger Battalion on 19 June 1942. The 1933 West Pointer screened over 2,000 volunteers from the only U.S. forces present in Northern Ireland at the time: the 1st Armored Division, the 34th Infantry Division, and V Corps units. He assiduously selected 520 men to begin training with British commandos in July 1942.[37] In a month and half, the battalion attached 50 rangers to two British commando battalions to gain combat experience in the disastrous Dieppe raid on 19 August 1942. Following the raid, Darby gleaned key lessons learned from the Dieppe survivors and fine-tuned his unit's assault techniques in Scotland. In late October, the 1st Ranger Battalion left Scotland and prepared for the TORCH invasion. In less than five months' time, Darby created a battalion out of nothing, trained it to assault fortified objectives at night, and participated in the TORCH invasion. Furthermore, despite his personal lack of combat experience, he displayed quality decision making under fire when he stopped the assault to use his preplanned mortar fires. Darby also had to command and control Dammer's element with minimal communication and no line of sight. Like almost any military operation, the rangers' assault suffered under the friction of war when key communications gear wound up on the bottom of the Mediterranean. The rangers benefited from a rigorous selection process to pluck quality men into their formation, which surely accounts for the raid's deftness. The battalion leaders—Darby and Dammer—utilized sound decision making under fire, which was not tied to the unusually high value of their subordinates.

Only Combat Command B (CCB) of the 1st Armored Division participated in the landings at Oran.[38] The rest of the division would follow in later convoys. The elements of CCB made up Colonel Paul Robinett's Task Force Green and General Oliver's Task Force Red. The Task Force Red "flying column" under the command of Lieutenant Colonel John Waters consisted of 1st Battalion, 1st Armored Regiment (less Company C), Company E, 6th Armored Infantry, one heavy platoon of Company B, 701st Tank Destroyer Battalion (75mm mounted on half-tracks), one platoon of the 16th Armored Engineer Battalion, and a reconnaissance section. The Task Force Green "flying column," led by Lieutenant Colonel John Todd, was smaller and made up by the 1st Battalion, 13th Armored Regiment (less Companies A and B), Company B, 6th Armored Infantry, a platoon of tank destroyers, and a platoon of armored engineers. Task Force Red landed upon the beach at the Gulf of Arzew (Beach Z) to the east of the city of Oran, while Task Force Green landed to the west of Oran at a small cove called Mersa bou Zedjar (Beach X). Both task forces enveloped Oran in an outer ring to seize airfields south of the city and to isolate it while the 1st Infantry Division attacked Oran from both sides, making up the inner ring on the assault. Essentially, the operation was a double, double envelopment with the 1st Infantry Division on the inside and CCB on the outside with armored columns. These armored columns, known as "flying columns," actually seized Oran before the 1st Infantry Division, occupied airfields that the paratroopers of Operation VILLAIN were supposed to hold, and defeated French armored attacks.[39] While critical to the success of the operation, the decision to outfit the tank battalions with light tanks would haunt the operations to come in Tunisia up until Christmas 1943.[40]

The limitations of the tank landing ships (LSTs)[41] forced CCB to outfit the two tank battalions, 1st Battalion, 1st Armored Regiment, and 1st Battalion, 13th Armored Regiment, with M3A1 light tanks. The Royal Navy identified the requirement for a large beaching ship that had the capabilities to deliver tanks and artillery quickly onto a beach. The low-profile tankers of the oil industry based in Lake Maracaibo, Venezuela, suggested the design to solve this amphibious need. The British converted three of these shallow-draft oilers ("maracaibos") to carry light tanks, artillery, and heavy trucks. Later assessments would confirm the essential role the LSTs would play in the success of the Oran landings: "The decision to load the 'maracaibos' to the maximum proved sound, because practically the whole operation, with the exception of the final assault on Oran, was conducted with the units loaded on the 'maracaibos.' Unloading of other units was so slow that their effect on the operation was negligible."[42] These predecessors to the LSTs did not have the capacity to transport M3 medium tanks (Grants) as they were about 3 inches too narrow to accommodate these tanks.[43] This limitation set the stage for the M3A1 Stuart to take the lead in all U.S. armor engagements up to the battles of Kasserine Pass. Armored regiments consisted of two medium battalions and one light tank battalion at this time, so these two light battalions, one from each armored regiment of the 1st Armored Division, received the mission to participate in the TORCH landings. We can assume that if the LSTs could have transported medium tanks, then planners would have selected medium tank battalions for the landings. No hard evidence exists for this assumption except for the fact that Allied forces did attempt to load medium tanks onto the oil tankers, but discovered that they were too large. While the Stuart had enough armor and firepower to handle any equipment the French could field along the coast of Algeria, the two tank battalions that made the landings had to keep using their under-armored vehicles (maximum of 1.75 inches on the front) with an under-powered 37mm gun and a small 54-gallon gas tank against the Axis formations they met in Tunisia.[44]

Waters at the TORCH Landings

Lieutenant Colonel John Waters received the briefing for the TORCH landings in London, and his orders forbade him to divulge any aspect of the plan to his tank battalion, 1st Battalion, 1st Armored Regiment, until the convoy left Clyde bound for Oran on 26 October 1942. The tankers became acquainted with the plan and viewed photos of the beaches while afloat on the LSTs and transports. With much anticipation after an uneventful sea journey, the two LSTs, *Tasajera* and *Masoa*, approached Z Red beach near Arzew and beached about 300 feet from the shore at 0345 on 8 November 1942. The 3rd Battalion, 16th Infantry, of the 1st Infantry Division, secured the beach and marked it for the tank column only forty-five minutes later. The engineers emplaced ramps for the vehicles rolling out of the LSTs, and they covered about 100 feet in an hour. Five hours later, Waters led his assembled column of tanks, infantry, engineers, and tank destroyers off the beach and raced for the Tafaraoui airport 35 miles away.[45]

By 1100, Waters's force of two tank companies and the reconnaissance platoon enveloped the airfield, lightly held by seven 75mm and 37mm antiaircraft artillery (AAA), guns originally misidentified as 88mm guns. Fortunately for Company B, approaching from the east, the AAA guns could not depress enough to fire due to the formidable parapets thrown up around the guns. Company A's assault from the south easily dispatched the hapless guns and their crews while French planes scrambled into the air to strafe Waters's forces, "to no avail." The seizure triggered two important events: first, possession of the airfield triggered the launch of the 309th Fighter Squadron (Spitfires) from Gibraltar to land on the captured aerodrome. Second, the Task Force Green column, commanded by Lieutenant Colonel John Todd, was diverted to seize the La Senia airfield instead of assisting Waters at Tafaraoui. Todd's task force did not seize La Senia until the next day, but by the end of the first day of the invasion, U.S. forces had established a working airfield at Tafaraoui. Waters and his men displayed a sound grasp of armored tactics through their rapid movement and maneuver against French guns and airplanes.

While advancing toward the Tafaraoui airfield, Waters had deftly deployed reconnaissance units along the major avenues of approach to provide early warning of French counterattacks. His preparedness paid off when one of his lieutenants, William Beckett, spotted fourteen French E35 Renault tanks in the vicinity of the village St. Lucien about 10 miles southeast of the airfield. In a hurry to reinforce Todd's assault on La Senia airfield and assist in the envelopment of Oran, Waters sent only his Company B under the command of Captain William Tuck to dispatch the armored threat, while he lead the rest of his force to La Senia. Tuck's seventeen tanks and four tank destroyers motored to St. Lucien and discovered the Renaults astride a small hill camouflaged among a copse of trees. Immediately, the commander used his tank destroyers to establish a static base of fire in a support-by-fire position 800 yards from the French tanks while he led his tank company in an assault to the right flank of the inferior French force.[46] Once the assault force sped into action, Tuck realized that he had not called for any artillery support or smoke; he would later say: "Boy if those guns had been effective, they would have wiped us all out. I never did that again." Despite these oversights, his unit had employed sound fire and maneuver and "crucified those tanks" with the temporary loss of only one Stuart and one U.S. sergeant in exchange.[47] Given the inferior firepower and armor of the Renaults as well as the unlikelihood of receiving artillery support for this subsidiary mission, Company B executed its attack well.

The first significant armored engagement of the TORCH landings resulted in an overwhelming Allied victory because of superior equipment combined with sound tactics. According to Tuck, the French company commander in this engagement went on to command a Free French tank company equipped with U.S. tanks and skillfully fought the German forces in the battles of Tunisia, displaying competence throughout the campaign.[48] He had occupied dominating terrain and employed camouflage to present a hasty defense against the Americans even though Captain Tuck's advantages overwhelmed the French. Tuck's grasp of armored doctrine is reflected in the following definition of fire and maneuver:

> The principal means of attack for armored force units is a combination of fire and movement to reduce enemy opposition. By fire and movement is meant the advance by certain elements or

> units (maneuvering element) protected by the fire of other elements or units (base of fire). The object of this form of attack is to advance elements or units to positions from which stationary fire will annihilate enemy opposition, or to a point where a swift assault over a short distance, strongly supported by fire, can be launched to overrun the enemy position.[49]

This concept is the basis for all maneuver ground warfare for armor and infantry units. The utilization of close air support (CAS), artillery (smoke and high explosive), naval gunfire, tank destroyers, and reconnaissance units facilitates the basic concept of fire and maneuver. Whenever possible, as Captain Tuck's thoughts suggest, maneuver units tried to employ these aforementioned techniques to increase their survivability and chances of success in offensive operations.

While Company B finished off the French tanks, Waters and his company commanders displayed their grasp of armored doctrine in their assault on the La Senia airfield. Waters's 1st Battalion conducted an assault on the airport from the north while Todd's tank battalion attacked from the south. Converging friendly forces on a battlefield requires heightened situational awareness and coordination among commanders. Throughout the capture of the airfield, Waters's units destroyed 75mm AAA gun positions and captured a full battery of dual-purpose 75mm AAA/AT (antitank) guns without receiving or inflicting any friendly fire casualties upon one another or Lieutenant Colonel Todd's units. Both armored battalions conducted their operations under heavy artillery fire and enemy air support, and against numerous roadblocks and gun positions manned by approximately 1,000 determined French soldiers. Both armored battalions had succeeded in the isolation of Oran, the destruction of counterattacks, and the seizure of airfields in two days. General Eisenhower's liaison officer for the Center Task Force wrote later about both Todd and Waters: "Both of these light tank forces were ably led and did splendid work."[50]

On the 10th, after the hectic fighting of the two previous days, Waters led his battalion into Oran with little trouble and noted that "the resistance and morale of the defenders was definitely broken."[51] In a display of insubordination fitting of his father-in-law, General George S. Patton, Waters searched in vain for General Allen, commander of the 1st Infantry Division, to brag that he was the first into Oran and he wanted to offer the general a ride on his tank.[52] After a close inspection of the city, Waters exited the city, passing some of his own tanks that had run out of gasoline, and found the corps commander, Lloyd Fredendall, on the outskirts. The 1st Battalion had not refueled for two days after leaving the airfield at Tafaraoui, and Todd's unit had outrun its communications (and made no attempt to reestablish them) with Task Force Green headquarters and was unable to get badly needed gas or ammunition in a timely manner. This nonchalant attitude toward logistics would haunt the armored forces across North Africa.[53] Fredendall climbed into the turret of the Stuart tank, and Waters took him into the city for a tour. Waters commented on the corps commander's demeanor: he was "afraid of being shot. He wouldn't stand up in the turret and look around. There wasn't any danger of getting shot. The people were welcoming us with open arms. He would only peek over the turret."[54] Fredendall's excessive personal caution would have disastrous effects later in North Africa. Adding fuel to fire, Waters led his battalion out of Oran back to the airfield once Fredendall had secured a negotiated peace. Apparently, Waters had gotten wind of a wharf loaded with captured Spanish red wine while exploring Oran and sent some of his trucks back to the port to secure the wine. He ordered his soldiers to line up and fill their helmets full of wine. "For several days they lay on the ground, just passed out from the red wine they drank."[55]

Courage and discipline, two intangible qualities that can make the difference on the battlefield, were lacking that day in Oran. However, the tankers of both flying columns had readily displayed these two qualities, along with their leaders' sound grasp of doctrine and combat leadership earlier in the landings. Given Waters's background and his recent as well as future exploits in North Africa and as a POW, his failure to utilize this time in a more constructive manner is puzzling, but was not uncommon among the U.S. Army of Operation TORCH.[56] This celebratory behavior ended abruptly and did not seem to negatively affect the future performance of the 1st Battalion. The ability of a unit to engage in such raucous behavior and still exude competence and calmness under fire is a tribute to the leadership and discipline of this particular tank battalion. However, Fredendall's behavior only foreshadowed his behavior in later actions in and around Kasserine Pass. Prior to the battles of Kasserine Pass, Fredendall employed his only digging assets to build tunnels for his corps headquarters, which was more than 100 miles from his most forward elements. Furthermore, he never ventured near the forward edge of the battlefield, even when there was no enemy contact, to ascertain the battlefield situation.

The Big Red One at Oran

Combat Command B did not secure Oran alone. Two-thirds of the 1st Infantry Division landed on Z Beach after the 1st Rangers secured their objectives. Both Combat Teams 18, led by Colonel Frank Greer, and 16, led by Colonel Henry Cheadle, raced ashore to hold the eastern flank of the Center Task Force and attack west toward Oran. With Colonel Alexander Stark at the helm, CT 26 landed on Y Beach, west of Oran, to protect the western flank and assault along the coast to Oran. Working with both Task Forces Red and Green, Terry Allen deftly synchronized his three combat teams to overcome, and when necessary bypass, French resistance, to secure Oran in three days.[57]

About a month after the TORCH landings, each participating division developed lessons-learned documents that Allied Force Headquarters collated into one staff memorandum. In a very thorough document, each war fighting function from medics to commanders addressed significant issues as well as the mundane, such as individual venereal prophylactic kits. The most glaring oversight—labeled "the biggest defect"—was the exclusion of battalion commanders from pre-invasion planning.[58] Only regimental commanders and higher received the TORCH brief before leaving England. On 5 November 1942, battalion commanders received sealed envelopes with their assigned missions and an overview of the landings. Once battalion commanders and their staffs obtained the invasion information, they were still isolated from their higher headquarters, which were usually on a different ship.[59] In less than three days, battalions had to brief their organizations, issue orders, and try to rehearse their mission.

The nine infantry battalion commanders of the three regimental combat teams managed to surmount these hindrances throughout their three days enveloping Oran. They executed division orders quickly and efficiently. Furthermore, commanders such as John Bowen, the commander of the 3rd Battalion, 26th Infantry, integrated all available combat arms. After an uneventful landing at Beach Y, Bowen and his men employed their engineer battalion mounted in M3 half-tracks with .50 caliber machine guns, regimental cannon company fire, and artillery to seize the town of Bou Sfer.[60] Likewise, Fred Gibb, 3rd Battalion, 16th Infantry Regiment, commander, isolated French artillery and machine-gun strongpoints with cannon company direct fire and indirect fire to maneuver unhindered south of St. Cloud into Oran.[61] Most resistance, such as snipers, artillery, and machine-gun fire, emanated from fortified strongpoints. To defeat these French units, battalion commanders synchronized combat functions without relying solely on their organic assets. In order to retain mobility, they knew from their first contact that they must defeat or bypass strongpoints utilizing all available means.[62] Fighting outnumbered, Lieutenant Colonel William Cunningham and only two companies of his 1st Battalion, 16th Infantry Regiment, defeated an entire Algerian regiment. Cunningham's rump unit was ambushed while they secured La Macta to protect the eastern flank of the task force. With an artillery battery in support and naval gunfire, the 1st Battalion secured the town and key bridges.[63] Routinely, commanders exercised initiative, such as Major Joseph Crawford when he led his 2nd Battalion, 16th Infantry Regiment, which was serving as the Center Task Force reserve, ashore without orders.[64] The division put them to work immediately to support an assault on the town of Fleurus. Although the Big Red One surpassed its sister units in amphibious training, it was what it accomplished once it got ashore that mattered, since most of its landings were unopposed. Despite the landing chaos and first combat experience, the commanders orchestrated combined arms warfare with an abundance of initiative.

Throughout their first three days of combat on the Algerian coast, the 1st Infantry Division's battalion leaders executed their missions with minimal guidance. The continual pressure to whittle down division and regimental operations orders to the bone during the GHQ maneuvers had paid off. The battalion leaders had practiced maneuvering their formations southeast across the United States with similar orders just over a year ago. The field order in this section is typical of the three issued by Terry Allen as his division rushed to close the vise around Oran.[65] Short fragmentary orders like this one rely extensively on subordinate units' familiarity with standing operating procedures. For example, all 1st Infantry Division combat teams knew what exact enemy activity to report to division headquarters based on the primary TORCH field order and previous exercises.[66] Most importantly, battalion commanders must also be comfortable with vague intent. All nine battalion commanders participated in the GHQ maneuvers and understood this. Brief orders transmitted Allen's intent quickly, but encouraged, and expected, commanders to prosecute their orders with their own initiative. Incompetent battalion leadership could not have executed such skeletal orders. Allen's mission-oriented guidance mirrored the *Auftragstaktik* espoused in the German operations manual, *Truppenführung*, which accounted for much of the Wehrmacht's success until the summer of 1941.[67] Modern warfare demanded decentralization to the smallest units as rapidly as possible. The officer candidate

"Hq 1st Inf Div
RENAN
2210, Nov 9, 1942

"FO #3

1. Omitted.

2. Div atks at 0715 Nov 1942 (See operation map scheme maneuvers and time of atk). CC B atks from S at 0730 in conjunction with 1st Div.

3. a. CT 18 see operation map
1st Bn CT 18 follows CT 18 after mopping up around ST CLOUD

b. CT 16 less 1st Bn, see operation map.
1st Bn CT 16 (brought forward in trucks follows in Div res).

x. Civilian snipers caught red-handed will be summarily shot.

Nothing in Hell must delay or stop this atk.

4. Attached.

5. Div CP initially follows 16th Inf.

ALLEN
Maj Gen "

First Infantry Division Field Order. This is a copy of a field order issued by the 1st Infantry Division shortly after the TORCH landings at Oran. It is very brief and authorizes subordinate commands to exercise wide latitude in its execution. "Development of Army Operational Technique," Memorandum for the President from Army Chief of Staff, 27 March 1943, Box 164, Folder 1, FDRL.

school graduates who filled the vast majority of company-grade positions became conditioned to the abridged orders, but their battalion leaders were ready before they disembarked for North Africa. The U.S. Army could not have implemented abbreviated orders down to the battalion level without the seasoned leaders who populated those positions.

Furthermore, the battalion commanders leading the units in Allen's combat teams were a product of a professional interwar army officer corps. Prior to the Oran envelopment, their superiors had judged their performance with efficiency reports that identified them as competent and motivated. Big Red One battalion commanders such as John Bowen, Joseph Crawford, Fred Gibb, and William Cunningham did not inherit their positions simply because of their regular army status with the onset of World War II mobilization. An honest evaluation system had ensured that these officers served in key leadership billets at the right time. If these Big Red One leaders had received inferior reports, then they most likely would have ended up at training installations or placed on higher-level staffs with lower levels of responsibility instead of closing the ring around Oran.

The Training of One Division: 1st Armored Division

The training regimen of the 1st Armored Division illuminates how battalion leadership prepared for combat and addresses the issue of "greenness" among U.S. soldiers that so many derided them for in the failures to come in February 1943.[68] The all-encompassing term "training" can refer to individual soldier tasks such as driving a tank, to crew-level tasks such as a tank firing, to unit-level tasks such as a tank company practicing a prepared defense. There are a thousand other actions in between these three examples that soldiers and leaders can practice that would fall under the heading of training. Quantifying how well trained a soldier or a unit is presents an almost impossible task, especially across an entire armored division. When soldiers train, they experiment with doctrine, learn lessons, hone standing operating procedures, and improve proficiency in their assigned duties and responsibilities; soldiers execute training for one purpose: to perform well in combat. Unfortunately, training cannot mimic all the complexities faced or terror shouldered by a person on a battlefield, but leaders wrestle to replicate the dynamic nature of warfare in their training. The knowledge that someone is actually trying to take your life does not exist in a training exercise. Given the limitations of training, what actions did leaders in the 1st Armored Division take to prepare their men and themselves for battle?

Four components of training prepared the leaders and soldiers of the 1st Armored Division to fight in North Africa: individual soldier training conducted at Fort Knox, pre-invasion exercises in Ireland and Britain, participation in the Louisiana and Carolina maneuvers, and field expedient instruction completed while in North Africa. These four aspects bear directly upon the 1st Armored Division's preparedness, but other factors would most certainly contribute to the development of soldiers and leaders. Personal experiences of the officers involved in such operations as the Mexican Punitive Expedition or tours of duty in the Philippines certainly had a pedagogical impact upon participants. The four components mentioned above address training that occurred within a three-year period to prepare the units that would actually fight in North Africa.

The Armored Force Replacement Center came into being in April 1941 to train and provide soldiers for armored regiments and tank battalions. Essentially, this center served as the basic training and military operational specialty station for U.S. Army armored forces. Equipment shortages and undeveloped policies plagued the training center throughout 1941. The fact that the center did not possess a single tank for the first few months and had to borrow them from the 1st Armored Division highlights these deficiencies. However, the center did develop into an effective school, but not until the late summer of 1942, when Major General Charles Scott took command. His experiences in forming the 13th Armored Cavalry Regiment as commander of the 2nd Armored Division and five months as an observer in North Africa with the British altered the regimen for the soldiers. Scott continually emphasized discipline and teamwork as well as a program that ensured all members of a tank crew could be interchangeable and fight under simulated combat conditions. Some of these soldiers who trained under Scott's new system would reach the units of the 1st Armored Division in time for the North African campaign, but most would not. Unit training would have to balance this deficiency.[69] Nonetheless, tankers who fought through Kasserine Pass spoke highly of their basic training at Fort Knox: "The gunnery instruction they gave us in the states was good. No, sir, I wouldn't change it"; and "the basic training they had in the states means a lot more to our boys over here. Every time they hit the ground you'll find them digging a helluva big hole."[70]

The entire 1st Armored Division participated in the well-known Louisiana and North Carolina maneuvers in 1941. Many of the participants left the division to attend officer candidate schools or to form cadres for other newly formed armored units and divisions. Lieutenant Colonel Waters's experiences showcase the fluid nature of personnel replacement at this time. He led a cavalry troop in the 2nd Armored Division in both maneuver exercises, but ended up as a battalion commander in the 1st Armored Division for Operation TORCH. Waters would later say that these exercises were "the most bitter, roughest training I've ever had, those Louisiana and Carolina maneuvers."[71] Waters possessed a horse cavalry background like his peers of the 1st Armored Division, Lieutenant Colonel Hamilton Howze and Lieutenant Colonel James Alger, though he did not find his experience in equine transportation an impediment to the transition to mechanized warfare. The maneuvers helped to facilitate this transition and prompted Waters to comment: "We've got a chance to go to war with something that's worthwhile and not a horse to fight the enemy armor."[72] The officers and soldiers seemed to have gained worthwhile experience from the maneuvers, but these experiences remained individual in nature since no unit would go to war intact with the same men it had at the end of the Louisiana and Carolina exercises.[73]

Prior to the division's deployment to Northern Ireland in May 1942, Colonel Paul M. Robinett became the commander of the division's 13th Armored Regiment. He left the best record chronicling the training programs instituted throughout the summer and early fall of 1942 in preparation for the TORCH landings. Even the casual observer knows that the terrain and climate of Ireland are nothing like those found in North Africa. Ideally, a unit's ability to train in the same climate and terrain that it will fight in would produce better-prepared soldiers and equipment; however, armies rarely have this luxury. Despite the compartmentalization of the Irish countryside, the 13th Armored Regiment formed a noncommissioned officers and an officers school to improve leadership, tactical and technical skills, and discipline. The regiment regularly used a 3,300-yard moving target range at Ballykinler, conducted rigorous night training, placed a constant emphasis on communication, forced all headquarters to participate in command-post exercises, and deployed on maneuvers as much as possible to improve doctrinal understanding and establish standing operating procedures such as tank stowage, methods of unit identification, and traffic control. Robinett said, "These exercises were thorough tests of physical fitness and tactical ability and were much more realistic than any I observed in the United States."[74] Other commanders, such as Waters, thought the maneuvers and ranges restricted, but felt that they trained their soldiers to "learn to shoot, to communicate, and to move."[75] Lieutenant Colonel Howze's memories of the training as the division training officer reflect a man frustrated by the restrictions of the countryside and fights about training land with landowners who did not seem to care if the Germans won or lost the war. Howze showed disappointment since his staff was limited to command-post exercises while training at the company level and below flourished, but the unit did receive brilliant after action reviews of their exercises from British officers with combat experience in France and North Africa.[76] The rock walls and bogs of Ireland prevented large-scale maneuvers, which no doubt denied the higher-level staffs the opportunity to flex their muscles. However, the focus on small units would seem to have bolstered doctrinal understanding at the platoon, company, and even battalion levels of command. Despite the benefits of all this training and drawbacks of limited space, there were bigger problems and oversights.

Three issues stand out as the most glaring faults: no exercises in coordinating close air support, very few training exercises that included tank and infantry cooperation, and unsatisfactory amphibious training.[77] Some historians, and officers such as Robinett, pointed to the command and control problems presented by the use of combat commands as well. Robinett's 13th Armored Regiment was subordinate to Combat Command B, led by Brigadier General Lunsford Oliver at the time, and as a regimental commander, Robinett did not approve of his staff being absorbed or used by Oliver while he became a "fifth wheel." Certainly, the two types of headquarters—the regiment and the combat command—created confusion. However, the combat commands were the functional commands that operated as combined task forces in combat. The regimental headquarters served mainly as an administrative, training, and maintenance headquarters. The division would use as many as four combat commands in Tunisia at one time to meet the enemy situation. These ad hoc organizations occurred throughout the war on both the Allied and Axis sides. Typically, leaders used the combined nature of the combat command to great effect; none of the Allied tactical defeats stemmed from these organizational issues. An unfortunate manpower requirement did affect the division in a negative way: ranger and paratroop units plucked volunteers from the 1st and 13th Armored Regiment formations while in Ireland. Senior officers and the official history of the division attest to the fact that it had little effect on the combat readiness since the replacements for the volunteers were trained by the "very best instructors" before being funneled into their units, but the loss of the motivated volunteers might have degraded the unit's readiness or morale.[78]

The last phase of the division's training occurred across the coastal plains and deserts of North Africa. This opportunity training could commence whenever time became available for a commander to take advantage of it. Mission requirements and the enemy situation hampered the thoroughness and volume of this training, but it occurred nonetheless on the terrain the division was fighting upon. Colonel Peter Hains, commander of the 1st Armored Regiment, later remembered after his regiment received M4A1 Sherman tanks in Britain just before they embarked for Tunisia: "We had been issued tanks that we had never seen before. And as we approached the end of our march, I sent recon [reconnaissance] up the mountain and made sure there was no people there. I just took each troop as they went by, 'Back off and fire as many rounds as you want and get familiar with your gun.'"[79]

His superiors had offered Hains the option of retaining the M3 Grants they had trained on in England or drawing the Shermans, which they would not get the chance to shoot or maneuver before they reached North Africa. Without giving it a second thought, Hains jumped at the chance to possess a tank with a rotating turret and increased armor protection: "You must be kidding somebody, we'll shoot it but we won't have any practice with it."[80] Not only does this incident show the ability and desire to train in the combat zone, but it also shows the mentality of willingness to accept new and superior equipment despite a lack of familiarity. Hains and all of his men had not yet fought against the Germans, and they did not make the equipment swap based on fear, but upon a willingness to adapt to changing equipment and circumstances.

Besides this example of opportunity training, the division took advantage of periods of bad weather or lulls in offensive operations to conduct training in Tunisia. Two key periods of this training were from about Christmas 1942 until the end of January 1943 and two weeks after the Kasserine Pass battles ended on 22 February 1943. This time was often used to field and train on new equipment like the Sherman or bazooka, but when the maintenance, weather, and tactical situation cooperated, the units of the 1st Armored practiced their maneuver tactics and techniques.[81] Given the circumstances and the rapid deployment of the division, its commanders usually made the best of their training opportunities, although Lieutenant Colonel Waters's wine festival is one black spot upon this record. Ironically, the 1st Armored

Division had the dubious distinction of being the only mechanized division to not take part in desert training, yet it became the only armored division to fight in the desert. *FM 17-33* (1941) states this about desert operations for armor battalions: "Special training and special equipment are required for desert operation. Prior to being moved to such a theater of operation, units should be given intensive training in a similar climate, on similar terrain, and under conditions that as nearly as practicable simulate conditions to be found in that theater."[82] The soldiers of the 1st Armored Division would not get perfect training or adequate training in amphibious landings and close air support, but the rapid nature of TORCH prevented a thorough training program in either Ireland or the Desert Training Center in California.[83] Furthermore, the deployment of desert-trained forces would have alerted Axis intelligence sources to the objectives of Operation TORCH.

The elements that made up both Task Forces Green and Red, known as Combat Command B, soon left Algeria and would have to test the depth of their training, leadership, and equipment against the Germans and Italians waiting for them in Tunisia. Only Combat Command B of the 1st Armored Division participated in the Oran landings and the race for Tunis that followed. The remainder of the division came ashore in mid-December. They would use this time to train and prepare while their brothers-in-arms furthered their combat education.

The Eastern Assault Force at Algiers

Similar to the Operation RESERVIST Oran port assault, the Eastern Task Force executed its own port attack against Algiers to prevent French sabotage against the port facilities.[84] Operation TERMINAL fell to the 3rd Battalion of the 135th Infantry, 34th Infantry Division, led by Lieutenant Colonel Edwin T. Swenson. His unit's preparation showcases the overwhelming challenges that prevented a reasonable chance of mission success. Of all the American units stumbling through the North African surf, the 3rd Battalion had to overcome the most glaring readiness deficiencies. The division lost approximately 260 of its most motivated soldiers to Darby's 1st Ranger Battalion's recruiting efforts five months prior to the landings.[85] It was the last National Guard division mobilized—created out of an amalgamation of Iowa and Minnesota regiments—and unlike the other TORCH divisions, its rapid deployment to Britain prevented participation in the GHQ maneuvers. As the only U.S. soldier labor pool in Britain in January 1942, the 34th Division spent

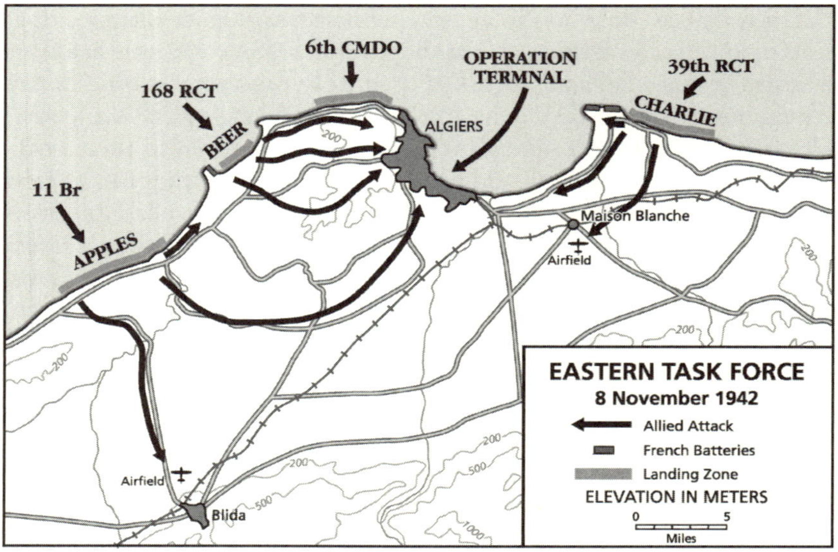

Map of the Eastern Task Force in Operation TORCH. From *Algeria–French Morocco*, The U.S. Army Campaigns of World War II, Center of Military History Publication 72-11, no date, 24.

the majority of its time in Britain doing labor details and guarding headquarters.[86] The 3rd Battalion was supposed to stay behind in Britain.

In mid-October, British concerns over French port sabotage spawned their manufacture of the frontal assaults on the most significant ports in Algeria, Oran, and Algiers.[87] Planners tapped the previously unassigned 3rd Battalion, 135th Infantry Regiment, to assume the Algiers assignment. In a frantic week, the unit rehearsed specialized assault techniques and individual combat proficiencies.[88] After the unit completed this rudimentary training, it boarded a transport on 26 October to begin their journey to Algiers. The 3rd Battalion had to transfer to two destroyers (the HMS *Malcolm* and the *Broke*) to conduct the attack once it reached the Strait of Gibraltar. To accommodate their new tight quarters, the battalion eliminated all administrative, supply, and communications personnel and reduced its medical staff to three minus the chaplain. Most importantly, the battalion had only four .30 caliber machine guns and two 81mm mortars with very limited ammunition.[89] Once Swenson's men had clambered aboard the ships, he briefed the mission and the four courses of action to his men. The assault's success hinged on two assumptions. The British 1st and 6th

Commando battalions had to neutralize the three French batteries that flanked old Algiers. Along with the commando attack, the Eastern Assault Force would land two hours prior to the TERMINAL attack, drawing any French forces in the harbor's vicinity to the outskirts of town. The four courses of action specified the units' objectives should both or only one of ships enter the Algiers harbor, as well as evacuation signals if the destroyers had to vacate the harbor after they had landed the soldiers.[90] In just over a fortnight, Swenson had prepared his unit for one of the most difficult military missions: an opposed amphibious landing. At 0220 on 8 November, Lieutenant Colonel Swenson received word that the Eastern Assault Force and commando battalions had landed; the destroyers started their approach to Algiers harbor.

HMS *Malcolm* and *Broke* started for the Algiers harbor and after four attempts, *Broke* severed the boom that protected the entrance and *Malcolm* retired back out to sea after sustaining several artillery hits from the French batteries. Immediately, Lieutenant Colonel Swenson knew that he would have to execute his Plan B, under which the *Broke* force would secure all the objectives.[91] Under a fusillade of heavy machine-gun fire emanating from docked ships and warehouses, HMS *Broke*'s crew rapidly found an empty berth to disgorge its lethal cargo. Swenson's men climbed down over the ship's sides to seize their three primary objectives: the docks, oil facility, and power station. Once the 3rd Battalion had secured its objectives, it consolidated to move north and seize the *Malcolm* force sites. As the dawn broke, however, the surrounding batteries and snipers suppressed the infantrymen's ability to maneuver. The French artillery fire serrated the *Broke*'s hull with twenty-two direct hits, forcing the captain to vacate the port. When the *Broke* sounded the siren to signal the 3rd Battalion to board the ship, Swenson wisely ordered his men to stand fast. He decided that the Eastern Task Force must be close to the city, and any attempt to load the ship under enemy fire would result in senseless casualties.[92] The 3rd Battalion would hold its objectives until relieved. To complicate their situation, ten French tanks maneuvered on the American soldiers, spraying machine-gun and 37mm ammunition. With no antitank guns, no naval gunfire support, and the loss of reserve ammunition that recently had sunk with the *Broke*, Swenson and his men built a perimeter amid the warehouses. By 1230, the situation was hopeless. The artillery and tank assault had spread fires in and around the perimeter, constructed of wood pallets and lumber, and no other friendly forces had penetrated the city yet.[93] After seven hours, Swenson surrendered the *Broke* force.

The 3rd Battalion suffered 15 killed and 33 wounded while inflicting 70 killed and 101 wounded on the French defenders. Unlike in Oran, the seven-hour struggle in Algiers prevented the French from sabotaging any port facilities. The original assault plan suffered egregious oversights, but Swenson had articulated a plan that all his subordinate leaders knew well.[94] Under the stress of combat, Swenson displayed courage and cool decision-making capabilities. Two of his decisions stand out. First, as artillery impacted the docks, he and his soldiers must surely have wanted to board the *Broke* to escape. The battalion had rehearsed that if the destroyers blew their sirens three times, the infantrymen would reboard the ships to evacuate the harbor. Swenson countermanded this order in a rapid manner. Finally, no commander ever desires or wants to order his soldiers to surrender.[95] This act is usually viewed as complete mission failure. On the other hand, few armies possess the moral fortitude and fanaticism to fight to the last men if the mission seems hopeless. Swenson's surrender reflects sound judgment under duress. The 3rd Battalion overcame the severe limitations imposed by the mercurial planning and rapid implementation of TORCH to perform effectively in its first combat. "It is suggested that the direct assault by infantry alone should only be decided upon if it is considered that they have a reasonable chance of achieving complete surprise": observers confirmed the futility of the port assault that lacked sufficient surprise, artillery support, and obscuration.[96] Planners clearly understood that not every assault element had enough time to become proficient in just basic landings, let alone assaults against fortified ports.[97] Yet, battalion leaders such as Swenson carried out their missions with a comprehension of their commanders' intents, but not with the appropriate training, resources, or support. Fortunately, Swenson and his men were liberated two days later when the French surrendered.

The remainder of the Eastern Assault Force rapidly controlled the countryside and forced a quick surrender of Algiers. By 11 November 1942, the Allies owned their much-desired lodgment on the North Africa continent at all three primary sites. Amazingly, they accomplished this with less than three months' planning time. At the tactical level, units often had far less time to prepare their units for first contact. Unlike their marine brethren, the army lagged far behind in effective amphibious operation implementation. More operational shortcomings awaited in Tunisia. When given the appropriate time and resources, leaders such as Darby and Waters performed well. Others, such as Raff and Swenson, suffered under overwhelming constraints. Regardless of mission failure or success, prewar training and conditions produced leaders who could execute missions with minimal guidance, and they could train their ranks of citizen-soldiers. They were also a product of a healthy officer corps, as evidenced by their efficiency reports. While the French were generally looking for a reason to surrender, the same cannot be said about the Italians and Germans hanging on for dear life throughout the Tunisian hills. To finish the job, battalion leaders still had to attack east to oust the Axis from North Africa for good.

4

The Race for Tunisia

Operation TORCH's success represented the beginning of the beginning of the Allies' struggles in the Mediterranean. A disjointed and, at times, spirited French resistance lasted only three days. The Vichy French surrender left the Allies far away from Axis forces to the east. Several operational and strategic factors conspired to hamstring the rapid beachhead exploitation. Weather, personnel management, logistical trials, command and control frameworks, intelligence, unit dispersion, and air support integration issues multiplied battlefield friction to cacophonous levels for maneuver battalions from the end of TORCH until late February 1943. These faults reinvigorated the Axis endeavor in North Africa despite the debilitating setbacks inflicted on the Afrika Corps by the British 8th Army at El Alamein just prior to the TORCH landings. Thus, as winter approached, the TORCH forces headed east into a reinforced Axis effort seeking to exploit opportunity in Tunisia.

Winter significantly affects the North African coast's weather in two ways. As the subtropical Atlantic Azores' system weakens, a continental high develops over northwest Africa, permitting frontal systems to migrate south over the coast. The fluctuation in highs encourages prevailing winds to shift from a northeasterly wind to a westerly direction. Wind changes and drifting highs rapidly increase the rainfall from November to February, with the most rain—3 to 4 inches—falling in January.[1] In mid-November 1942, the downpours began as the Azores' high weakened. Inches of rain precipitously degraded road and airfield quality.

Even if the weather had not conspired against the Allies, the road and rail network could not sustain the offensive to the east. Most bridges were only wide enough to accommodate one-way movement, and many roads narrowed down to one-lane traffic due to soft or eroded shoulders. The rain further eroded the unpaved roads, made cross-country movement mostly impossible, and forced logisticians to turn to the rail network. Good news did not await them. From Algiers, the rail line ran to both Souk Ahras (near Tebessa) and Souk el Arba—the main railhead for the offensive. Over this 800-mile supply line, the Allies could count on only nine trains a day, and three of those hauled the bare minimum for civilian supply and power generation (coal) requirements. Therefore, under optimal conditions six trains a day could deliver 250 tons of military supplies and equipment to the railheads.[2] Unlike later in the war, soldiers did not enjoy an overwhelming logistical edge. Battalion commanders often had to demand simple items such as boots or develop innovative solutions.[3] In one instance, a battalion commander hired local Arab women to repair uniforms and create new ones for his soldiers.[4]

Replacement soldiers arrived at North African depots ill-prepared to integrate into combat units. Clearly, a quality problem existed between Army Ground Forces (AGF) soldiers and other military recruits such as Army Air Corps members. Underlying this entire question of recruit quality is the U.S. Army's system used to rate incoming soldiers. It had several shortcomings that did not benefit maneuver ground forces. The army classified all incoming personnel with the Army General Classification Test (AGCT) into five classes to determine each man's "ability to learn."[5] The Army Air Corps drew a huge proportion of Class I and II soldiers—the highest quality—to fill cockpits and maintenance crews to wage Operation POINTBLANK, the only offensive action initially directed against Fortress Europa. When these high-level recruits washed out of pilot training, the air corps integrated them into key ground-personnel positions. In comparison, combat divisions received 27.9 percent of Classes I and II, and the air forces got 41.7 percent.[6] The army evaluated recruits' physical, intellectual, and occupational skills, but it could not measure a man's potential as a warrior or combat leader. Its system benefited the Army Air Corps, Army Service Forces, and Army Specialized Training Program by filling their ranks with either highly intelligent or highly skilled people. In summary, combat divisions often did not know what type of soldier they were getting or how he would perform in combat. However, the average combat division soldier in 1942–1943 did not possess the same intellectual capacity as those in other branches.[7] The army also struggled to obtain motivated men for the infantry and armor. Little incentive existed for self-aware men to enlist in ground combat forces such as the infantry and armored forces. Even though the army permitted volunteers until the end of 1942 to select their own branch, only 5 percent chose infantry or the armored force, and the majority chose the air corps.[8] Finally, the army did a poor job in its competition for recruits when compared to the other services. The U.S. Navy and Marine Corps actively recruited higher-quality personnel. For example, the navy's aggressive recruitment of seventeen-year-olds denied recruits of high intelligence and good physical condition to the AGF. The combat soldier in maneuver battalions was not incompetent or cowardly. Based on AGCT ratings, however, unit officers and NCOs did not know exactly what they were getting, and they had to spend more energy training their men.[9]

Research and analysis have revealed that the AGF units eventually overcame this quality obstacle.[10] Early on, however, replacements flowing into North Africa were grossly undertrained in a thirteen-week program.[11] As the soldier progressed from depot to depot from the continental United States to North Africa, he languished unsupervised for months. Soldiers arrived "physically soft," riddled with poor discipline and atrophied skills.[12] These unprepared replacements required more training by their gaining units. The ground observer and commander reports were so damning that the AGF adopted a seventeen-week training regimen coupled with a more streamlined and organized system in August 1943 to inject the right replacement into the correct unit.[13]

A sea of Officer Candidate School (OCS) graduates filled approximately 80 percent of the company-grade billets in U.S. Army maneuver battalions. By most accounts, the army could not have functioned without them. Indeed, one infantry regimental commander commented:

> Let me say a word about these OCS people in case you have not had any contact with them. They are far in the way [*sic*] the best that I have seen in the Army, and for the job they have to do I had just as soon have them as any graduate of the Military Academy joining his first regiment. They are well grounded, interested in their job, industrious, ambitious, and on the ball twenty-four hours a day.[14]

The renowned Dick Winters, of *Band of Brothers* fame, and Joseph Dawson, credited with leading the first company off Omaha beach on D-Day, were both OCS graduates who proved themselves as combat leaders.[15] The Reserve Officer Training Corps and the battlefield promotion system also produced more than capable company-grade leaders such as Charles B. MacDonald and the gutsy Audie Murphy.[16] For all of their accomplishments, these critical officers were unprepared to step into battalion leadership positions at the war's outset. Dawson served as an assistant regimental S3 and aide with the 1st Infantry Division in North Africa, and Murphy was integrated as an enlisted replacement into the 3rd Infantry Division in February 1943 in Morocco. Both benefited from these experiences that they utilized later in the war. Even though a few men such as Winters would rise to command battalions, this occurred only after three years in the army and multiple months in combat. These officers provided the essential killer mentality, intellectual acumen, and leadership required for effective combat organizations, but had yet to learn battalion-level command and control systems and combined arms synchronization.[17] In North Africa, junior officers indicated a state of "confusion" when they had to integrate infantry, tanks, artillery, tank destroyers, and close air support to attack an enemy force.[18] Battalion leaders in the MTO relied heavily on OCS officers' contributions, but they also had to mentor and coach them to develop their skills.

Two traditional Allied strengths that helped to guarantee victory were absent in North Africa until April 1943: military intelligence and airpower. Neither function should receive credit alone for Allied victory, but the defeat of the Axis in Europe seems unlikely without the contributions of Ultra and the air dominance flaunted throughout the European skies. The Allies did not fully "break" the German Army's Enigma for the entire war until December 1942, and even then, they were unable to foresee the intense Axis reinforcement in Tunisia or the eventual offensives in February 1943.[19] Without any reliable strategic or operational intelligence from Ultra or air observation, battalion commanders were left to glean any enemy information only from patrols, indigenous French colonial forces and peoples, and observation.

Even though any combat arms officer who participated in interwar army unit or AGF maneuvers grasped the dire need for better air-ground coordination for reconnaissance and close air support (CAS), air support arrived stillborn in North Africa.[20] The ground forces remained tied to event-driven operations, and the air corps stayed bound to time-driven missions; never did the two attempt to meet until after the Kasserine Pass battles in February 1943. Army staffs did not have properly trained intelligence cells to interpret photo reconnaissance, and ground and air staffs rarely coordinated their efforts.[21] It often took twenty-four hours or more for staffs to receive aerial photographs.[22] Ground units did not have HF or VHF radio sets to talk to pilots. Even if both arms had developed air-ground coordination teams, there were no planes to direct to ground targets. Ground and air headquarters often employed motorcycle messengers to deliver requests for air support that migrated ponderously up and down both command echelons and that were invariably executed long after a ground unit needed the aircraft.[23] One battalion executive officer laconically stated, "We have had no close support from our Air Corps."[24] The air corps remained focused on hitting operational targets such as shipping bases and even lacked any appropriate fighter-bombers to support ground units. For reasons of doctrine, airfield inadequacies, and weather, air corps airframes were never on call for troops in contact or planned offensives.[25] The U.S. military did not solve this quandary until mid-1944 even though the gross desynchronization between air and ground received plenty of attention in North Africa. Indeed, *FM 100-20, Command and Employment of Air Power*, was produced in July 1943 to rectify the situation, but served as only an evolutionary step. This conflict did not assist the battalion commander, who probably agreed with the following statement from *FM 100-20*: "Land forces operating without air superiority must take such extensive security measures against hostile air attack that their mobility and ability to defeat the enemy land forces are greatly reduced."[26] All of these numerous shortcomings had yet to metastasize in the wake of TORCH's success. The Allies hoped for the best and pounced on a chance to seize the operational initiative by severing the Axis lifeline in Tunisia.

In order to deny the Axis forces their logistical lifeline, the Allied plan called for the British 78th Infantry Division, supported by Combat Command B, 1st Armored Division, to make a rapid thrust east along three axes. The 36th Infantry Brigade Group would take the northern route supported by Company E, 13th Armored Regiment, CCB; the 11th Infantry Group would take the southern axis supported by the 2nd Battalion (minus company E), 13th Armored Regiment, the 175th Field Artillery Battalion, and Company C, 701st Tank Destroyer Battalion—all from CCB; and Waters's 1st Battalion, 1st Armored Regiment, would support the middle axis, called Blade Force, under the command of Colonel R. C. Hull. Prior to the launch of this ill-fated offensive, the Axis built up its forces in the Tunisian bridgehead at a blistering pace: airlifts brought in 15,000 men and 581 tons of supplies, and sea transport brought three fresh German divisions, two Italian divisions, 176 tanks (including a few examples of the new Mark VI Tiger),[27] 131 artillery pieces, 1,152 vehicles, and 13,000 tons of supplies to meet the Allied threat. As the German resources mounted, both sides scrambled to seize the routes leading into Tunis, especially in the Medjerda River valley. Poor weather, Axis reinforcements, nonexistent logistical support, insufficient forces to mount an offensive, and inferior

Allied equipment eventually resulted in the failure of this drive for the Tunis bridgehead. Tactical engagements by the Blade Force, CCB, 1st Armored Division, and 1st Infantry Division made up the bulk of the U.S. effort. These actions are telling of the previously outlined problems as well as how battalion commanders utilized doctrine and exhibited combat leadership. After subduing the French, they faced their first tests against the Germans and Italians in these fights.[28]

Battalion Leadership in the Fight for Chouigui Pass

"We did very well against the scrub team. Next week we hit German troops. Do not slack off in anything. When we make a showing against them you may congratulate yourselves," said Waters to his battalion after their exploits in and around Oran.[29] The 1st Battalion moved its tanks via light-rail to Souk Ahras to begin its movement toward Tunis. Allied headquarters selected Waters's unit for this mission since they were deemed ready and they had light tanks, which were the only type that could fit through the tunnels and bridges of the light-rail system. Given a very ill-defined mission to establish a tank-infested area in the Mateur–Bordy–Sidi Bou Said area on either side of Chouigui Pass and to reconnoiter the bridges at Tebourba and Djedeida, the battalion disembarked the railhead as an almost independent tank battalion with no artillery, maintenance, or infantry support, with air support in the form of one Canadian Spitfire squadron to cover the entire front, and with a supply line that stretched almost 60 miles.[30] The men under his command took Waters's words of warning to heart in the upcoming actions.

Lieutenant Colonel Waters identified the Chouigui Pass as the key terrain feature that his unit had to hold in order to infest the prescribed area. While securing this pass, the 1st Battalion endured constant enemy air attacks—ten in one day—but their .30 caliber machine guns proved woefully inadequate as antiaircraft weapons, and only the handful of .50 caliber machine guns in the unit had any effect on the German Me-109s and Stukas (Ju-87s). German infantry occupied a walled farm in view of the pass, but the 1st Battalion was unable to eliminate this threat despite determined attacks. During the attack on the farm by one tank company, one of the Stuart tanks became immobilized, and the tank commander left one of the crew members behind. Upon discovering that the man had been left behind, Major Bill Siglin, the company commander, and the tank commander of the missing man conducted a night raid in a jeep to rescue the wounded and stranded tanker. The only casualty was

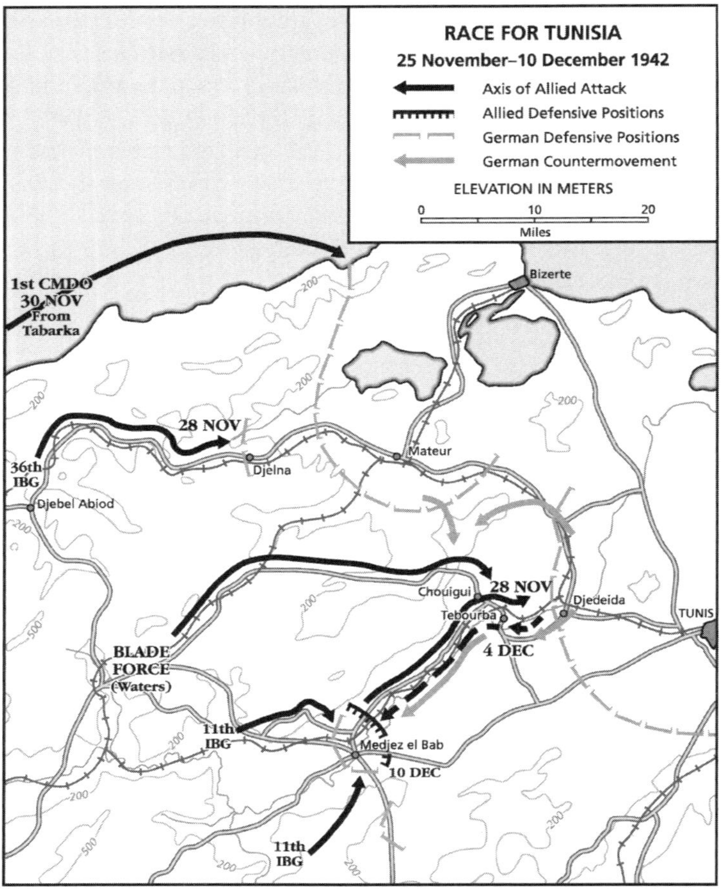

Race for Tunisia in late November and early December. Poor weather and German reinforcements foiled this rushed attempt to end the campaign in North Africa quickly. Notice the Blade Force axis of advance, which contained CCB, 1st Armored Division. From *Tunisia*, The U.S. Army Campaigns of World War II, Center of Military History Publication 72-12, no date, 9.

the jeep, which promptly died once it reached the company assembly area, riddled with machine-gun fire.[31] The lack of infantry, artillery, and air support for Blade Force prevented a potent combined arms integration in these two instances. The infancy of the logistical support within the theater of operations, weather, and the considerable distances covered by Blade Force created this unfortunate situation. The tankers of the 1st Battalion adapted to these limitations to remain calm under fire and even display bravery. They would soon try to even the odds and make up these shortages at the enemy's expense.

Waters launched Company C on a raid the day before Thanksgiving, the 24th, to reconnoiter the bridges across the Medjerda River near the towns of Tebourba and Djedeida. The company moved deliberately and destroyed at least a company's worth of German reconnaissance vehicles, including motorcycles and Volkswagen scout cars in the eastern side of Chouigui pass. Next, the company obliterated the small German guard detachment at the El Bathan bridge in Tebourba and continued moving east to Djedeida. As the company crested a ridge, the Djedeida landing ground was flush with parked Me-109s and Stukas lightly guarded by a light flak battery and easy pickings for the fast Stuart tank crews, smarting for revenge after so many air attacks.[32] The tank company advanced with two tank platoons abreast and one trailing in reserve. "The planes were fat geese on a small pond. The tanks blasted them with 37mm HE [high explosive], riddled them with canister, and set them afire with .30 caliber tracers. Several tanks enjoyed themselves in an orgy of destruction, physically crashing lines of parked aircraft by running over their tail assemblies."[33] Two planes managed to take off and killed two tank commanders, but the tanks won the day with the destruction of thirty-six planes, including Me-109s, Stukas, and Ju-88s. This engagement typified the armored concept of exploitation to "get behind the enemy" in order to spread panic and break the enemy's confidence.[34] The next day would showcase the unit's use of armored doctrine.

The 1st Battalion assumed the reserve mission for Blade Force and dug in on both sides of the Chouigui Pass, also known as "Chewy Gooey" to the tankers, to rest and refit. Waters deployed his forces so they could defend the pass and move to meet any threat that would try to come through the critical piece of terrain. His unit had replenished its fuel, ammunition, and rations, performed maintenance on weapons and vehicles, prepared foxholes, employed observation posts and air-alert guards, and selected fighting positions for their Stuarts. Company B occupied a reverse slope position on the east side of the pass, which offered a commanding view of the road through the pass at only 100 yards away; after selecting this position, the commander ensured that all his vehicles maintained camouflage and that their vulnerable hulls stayed behind the ridge. The soldiers of 1st Battalion had prepared for the worst and hoped for a relaxing Thanksgiving Day.[35]

Back at General Walther Nehring's German XC Corps headquarters, the level of anxiety increased from the reports of American tanks charging across German airfields, complicated by false sightings of U.S. units only 9 miles from Tunis. At the behest of Field Marshal Albert Kesselring, Nehring sent a reconnaissance force consisting of a company each from the 11th Parachute Engineer Battalion and the 3rd Tunis Field Battalion as well as a tank company from the 190th Panzer Battalion toward the Chouigui Pass to ascertain the Allied threat. The tank company had six Mark IVs with a high-velocity 75mm cannon and three Mark IIIs with 50mm guns. As the German column approached the fortified farm attacked by Waters's men earlier in the week, the American tankers spotted them.[36]

The Axis tanks detected movement in the pass and opened fire while the 1st Battalion, with their poor sights and binoculars, struggled to identify the vehicles. In order to deploy Company A, Waters ordered his three-gun assault section equipped with 75mm pack howitzers mounted on half-tracks to race across the valley floor and force the enemy tanks to react. The hail of 75mm fire got the attention of the Axis tankers, who replied with a heavy dose of AP (armor-piercing) ammunition. While many of Lieutenant Wacker's shots had scored hits on the tanks, the shells lacked HEAT (high-explosive antitank) capability that could have penetrated the thick hide of the German tanks. Fortunately for the assault platoon, the German tanks missed their marks, and Waters ordered Wacker's platoon to disengage. The assault gun crews shot some white phosphorus shells to provide a smokescreen and withdrew out of range.[37]

Lieutenant Colonel Waters had Major Tuck's Company B remain in hull-defilade positions while he ordered Company A to advance through the pass to attack the tank column on its right flank while Company B waited mutely on the left flank. Major Bill Siglin led Company A's twelve functioning tanks toward the right flank of the Germans. The blistering pace of operations and poor maintenance supply system had claimed five tanks due to mechanical failure. The Axis forces had deployed some light Italian Semovente tanks on their flanks, which Company A quickly dispatched. The German tanks barely had time to deploy off the road to point their hulls in the direction of the attacking Americans. As they focused their fire and attention on Company A, their superior armor provided protection while their lethal guns destroyed six of the light Stuart tanks. One platoon leader pumped eighteen 37mm rounds at one of the Mark IVs with no appreciable effect. Tuck's B Company seized its opportunity and let loose with their underpowered guns into the rear engine compartments of the German tanks. The 37mm guns were just powerful enough to render the tanks immobile, but they could not cause the German tanks to brew-up (fuel and ammunition exploding from the inside of a tank)since they were using ammunition largely used for training instead of armor-piercing ammunition, not yet available.[38] The Americans had destroyed at least nine enemy tanks and forced four others to retreat. Just as the tank battle ended, both Companies A and B sighted the trucks disgorging two companies of infantry in the vicinity of the fortified farm. Tankers unleashed an assault against the vulnerable infantrymen, and with machine-gun and tank fire they slaughtered them in the vineyards and olive-tree groves surrounding the farm. Next, they battered down the front gates of the farm with their tanks and killed the defending garrison. While machine-gun and tank fire raged across this battlefield, the battalion surgeon, chaplain, and medics cruised around to evacuate the wounded from the knocked-out tanks. The first American-German tank battle of the war had ended in a victory by U.S. tankers.[39]

These small engagements speak to the soundness of armored doctrine executed at the battalion level and below by competent combat leaders.[40] The 1st Battalion, 1st Armored Regiment, fought isolated in the early phases of the Blade Force offensive with inferior tanks, but overcame these shortcomings to execute impressive raids such as that on the Djedeida airfield, destroy enemy reconnaissance units, and conduct lethal defenses and counterattacks at Chouigui Pass. The soldiers exercised discipline in their calmness under air and ground fire, preparation of camouflage, and construction of foxholes. Despite logistical isolation, internal maintenance assets produced acceptable operational ready rates even with excessive mileage that averaged about 40 miles a day on tracked vehicles. The medical evacuation system reflects a unit that trained to do this and had a system to get wounded men off of the battlefield. The tactical decisions such as the use of defilade positions and of smoke for withdrawal show sound leadership. Although Major Siglin was killed in the Chouigui Pass fight, he had displayed courageous leadership, and so did the other two company commanders, Majors Tuck and Barlow, each of whom would later command the 1st Battalion, 1st Armored Regiment, during different periods of the Italian campaign. Not only had Waters maneuvered his force over a 100-square-mile area, but the leadership demonstrated by his subordinate officers, including company commanders and platoon leaders, indicates that he had trained, coached, and mentored these men into a skillful team.[41] Looking at this unit, one does not find "green" troopers groping to understand the complexities of armored warfare. Combat observer reports confirmed this through their constant praise of units' employment of tactics taught by the service schools.[42] Despite the technological inferiority of the Stuart, the 1st Battalion had achieved significant results, although this gap between light tanks and the heavier German tanks streaming through the ports of Tunis and Bizerte would widen, helping to quash the quick attempt to close the North African front.

The Blade Force moved to Tebourba, where it endured severe enemy air attacks and met German armored counterattacks until Colonel Hull withdrew his force on 4 December. From the 4th until the 26th of December, the Allies poured forces into Tunisia to build on the earlier successes of the offensive, while the Axis forces pushed back to protect their bridgehead. The entire Combat Command B gained combat experience during this time, with mixed results. The only medium tank battalion, 2nd Battalion, 13th Armored Regiment, languished under the leadership of Lieutenant Colonel Hyman Bruss. Colonel Robinett relieved him on 6 December, and Major Henry Gardiner, the executive officer, assumed command, having demonstrated competent leadership in the absence of Bruss's lackluster talents. Gardiner had taken command of two tank companies when Bruss failed to act to support 1st Battalion, 6th Infantry, at Medjez El Bab. His leadership prevented a total rout of the battalion.[43]

Medjez El Bab: Indicators of Weak Leadership

In early December 1942, logistic constraints forced the Axis to decide which front to pursue in earnest in North Africa. Generaloberst Juergen von Arnim's newly minted Fifth Panzer Army was directed to conduct an aggressive offensive to shore up the Tunisian bridgehead.[44] On 6 December, the 10th Panzer Division attacked the static Allies south along the Medjerda River valley toward the critical junction of Medjez El Bab. Combat Command B, 1st Armored Division, waited in the valley. The upcoming fight would carry on until 10 December and reveal what battalion leadership could and could not do against a determined German effort.

William Kern's 1st Battalion, 6th Armored Infantry Regiment, thinly defended the valley along a 5-mile exposed forward position with deep ravines separating many of the units. Kern anticipated the German attack after constant enemy air observation and registration of mortars and artillery on the 5th.[45] His report went unheeded by CCB headquarters. The Germans attacked in task force–sized elements that included a well-balanced mix of infantry, tanks, artillery, and antitank guns supported by coordinated close air support. Unlike the Germans, CCB kept its infantry and armor battalions *pure*, meaning that battalion leaders possessed only their organic combat arm. Kern later noted, "At no time did we have the opportunity to employ the battalion as Armored Infantry as taught by our Armored Force Schools. We were on the defensive all of the time that we were in contact. The only time that we have operated with tanks was in the landing operations at Oran."[46] Kern's isolated infantry units now faced German combat teams attempting to defeat them in detail. With only two artillery batteries and one assault gun in support, Kern deftly coordinated his infantry companies in a series of withdrawals and counterattacks to stave off disaster, while the CCB struggled to coordinate an armored counterattack.[47] Six hours after the first German assault against Kern's battalion, Bruss's 2nd Battalion of medium tanks arrived too late to offer any real assistance. Over the next four days, a series of costly armored counterattacks led by Waters and Todd and some gritty defensive stands by some engineers and French artillerymen convinced the Germans to call off the attack. Combat Command B received orders to withdraw from the valley to consolidate and reorganize. Led by the senior battalion commander, Lieutenant Colonel John R. McGinness, instead of the regimental commander, Robinett, or combat command commander, Oliver, the unit withdrew at night under enemy artillery fire.[48] As McGinness approached the bridge at Bordj Toum, a rumor of German tanks at the bridge swarmed through the ranks. McGinness believed the rumors. He ordered CCB to abandon the paved road for a dirt road that led to another bridge not defended by imaginary tanks.[49] Days of rain had severely degraded the route's trafficability. Almost two battalions' worth of equipment became inextricably mired, and McGinness gave the order to abandon the vehicles.

McGinness's leadership failure represents a prime example of ineffective battalion leadership. At age forty-five, he was at least ten years older than any of his peers. He had served as an enlisted soldier from 1918 to 1920 and entered West Point in 1921. After graduating in 1925, he returned to his enlisted unit at Fort Hays, Ohio. He attended the Infantry School in 1928 and, again, returned to Fort Hays. Unfortunately for him, McGinness missed the West Point and service school reforms of the 1930s. He spent his formative officer development years at the same post in Ohio, which did not challenge him or expand his knowledge. Peers only remembered him for how well his boots were shined.[50] John McGinness was Old Army through and through and unprepared for the rigors of battalion command in combat. His miring of an entire unit was scorned by peers and observers alike and was even briefed to the president.[51] His superiors relieved him on the spot, making him the second battalion commander (after Bruss) to get the ax in only five days. Oliver and Robinett should not escape scrutiny either.

Both the regimental commander and the combat command commander failed to task-organize their battalions for combat, and they did not shape the battlefield for their subordinate units. Kern's desperate fight reveals these shortcomings. His battalion did not possess any operational control over tanks or the ability to call for air support. Once Kern's flanks came under heavy pressure from German attacks led by tanks, his ability to defeat this onslaught was unlikely, given his task organization. CCB's answer was to send a tank battalion to assist Kern's 1st Battalion. This uninspired decision held the same perils for the counterattacking tank battalion. As the tank battalion approached Kern's positions, German forces simply set hasty defensive positions and defeated the force with antitank guns and tanks in static positions.[52] The tank battalion had no infantry to clear or assault the German fixed positions.[53] Maneuver battalion commanders could operate best when they owned tanks, infantry, engineers, and tank destroyers, but not field artillery batteries.[54] The combat command thought that the two field artillery batteries attached to Kern would augment his force enough. Field artillery, however, achieved more devastating results when kept together as a battalion. Typically, as was the case with Kern's fight, field-artillery batteries integrated with maneuver battalions often got overrun and became highly susceptible to counter-battery fire.[55] When field battalions remained pure, they could mass fires better and respond to critical missions dictated by the combat command.[56] Dividing artillery battalions to operate as individual batteries robbed the combat command commander of any method to shape the fight for subordinate units. Until the Allies perfected air-ground integration, massed artillery fires were the only option available to a combat command.

If the U.S. Army suffered under less than inspired tactical leadership at the regimental level in North Africa, it benefited tremendously from its indirect fire when used correctly. Interwar experimentation had perfected the development of the battalion fire direction center (FDC). This innovation coordinated all fires so maneuver commanders could mass a battalion or more of indirect fire on an intended objective. And artillery units usually got better and better. Unlike the withering casualties endured by infantry and armor battalions, artillery battalions took far fewer casualties.[57] Thus, they retained critical expertise to perfect observation techniques and communications, and to decrease mission time in FDCs. The Allies could never have stymied the German counterattack at Kasserine Pass or dislodged the Axis forces from their defensive positions in Tunisia without superior artillery fire.[58] Maneuver battalion commanders integrated this tactical advantage, and more often than not it was their only combat multiplier to achieve combined effects against their foes. Throughout the Medjerda River valley struggles, however, the artillery fire was more often than not dribbled away. In this race for Tunis, maneuver battalion commanders such as Kern, Todd, and Waters relied mostly on their own unit without massed artillery support.

Both Lieutenant Colonels Todd and Waters led their units well, but their tanks could not match up against the Germans. Combat Command B had lost 85 out of 108 M3 light tanks during the race for Tunis and 25 of 54 medium tanks due to enemy action. The inability to perform effective maintenance throughout this period of extremely high operational tempo contributed to high tank losses and a heavy wear-out rate as well.[59] Waters survived the onslaught of German armor, while Lieutenant Colonel Todd lost his life assisting 1st Infantry Division at Longstop Hill, but was posthumously awarded the Distinguished Service Cross.

The crews of the M3 Stuarts understood the deficiencies of their tanks, but they fought until Christmas with sound tactics. Inept leadership, not doctrine, had squandered the capabilities of the M3 Grant during the race for Tunis. Later in the campaign, Colonel Robinett had to choose whether to retain an M3 Grant–equipped battalion led by a known competent commander, Major Gardiner, or gain an M4 Sherman battalion led by an untested commander, Lieutenant Colonel James Alger; he chose the M3 battalion despite their disadvantages because he valued the leadership he knew and had trained himself over a technological edge.[60] After six weeks of combat and McGinness's foolhardy decision making, Combat Command B needed to refit. It was relieved in the Medjerda Valley by the 18th Infantry Regiment; its 1st Battalion's actions showcase light infantry combat in Tunisia.

The Struggle to Coordinate at Longstop Hill

The 1st Battalion, 18th Infantry Regiment, 1st Infantry Division's participation in the Battle of Longstop Hill illustrates the almost ludicrous command and control issues that beleaguered Allied efforts to seize Tunisia before Christmas.[61] In one last attempt to seize Tunisia before the Germans cemented their grip, the British 6th Armored Division and 78th Division planned an attack up the Medjerda River valley. Initially, the 18th Combat Team (CT 18) was hastily placed under the command of the 78th Division. The original plan called for the Coldstream Guards to seize Longstop Hill and 1st Battalion, 18th Infantry Regiment, would then relieve them and detach from CT 18 and come under control of the Guards Brigade. After the 5th Northamptonshire Regiment mined the Tebourba Gap between the mountain and the river, they were to fall under the command of CT 18 until 1800 hours on D-Day, when they were supposed to be relieved by the 1st Battalion of the French 4th Zouaves. The Northamptonshire Regiment would then come under command of the 11th Infantry Brigade, and the Zouaves were to fall under CT 18 at 2400 (midnight) on D-Day plus one. While all these task organizations transpired, CT 18 was supposed to conduct two separate night attacks. The plan was anything but simple, and further complicated by units from three different nations.[62]

In reality, the attack never progressed beyond the attack on Longstop Hill. The British Coldstream Guards seized only two of the five peaks, failed to guide Lieutenant Colonel Robert York's 1st Battalion into position, and vacated the hill at 0400 before both battalion commanders had conducted a proper handover. As the sun rose, giving the hill a purplish hue, the Germans counterattacked from the other peaks. They recognized how critical the road, rail, and river network was to defending Tunis, so they had launched a *Kampfgruppe* under the command of Col. Rudolf Lang, composed of elements from the 69th Panzergrenadier and 7th Panzer Regiments to brush aside the Allies.[63] York's men initially blunted the German counterattack and, for a brief time, controlled all five peaks that comprised Longstop Hill, all without artillery support. The British Coldstreams' artillery colonel believed the artillery was not yet "competent" due to its recent arrival and needed more time to set up.[64] During the attack, York lost almost his entire A Company to a well-placed German ambush as they maneuvered west.[65] Lang turned up the heat and committed his entire command to push York and his battalion back to its original positions by late afternoon on 23 December. Things continued to get worse. The Coldstream Guards responded to York's request for support to assist his isolated command. The Guards attacked through York's position to control the entire hill once again.[66] Throughout the night, the Allies prepared positions to defend the hill against the assured German counterattack that would come in the morning. As the Germans enveloped the hill in the morning, two infantry companies, one British and the other one French, that tied into York's right and left flanks withdrew under German fire without notifying him.[67] Enemy artillery fire cut all wire communications, and a soaking rain denied the use of most wireless radios. To compound this issue, forward observers with the only working SCR 536 radio were out of position, which delayed any indirect fire for the first two hours of the German counterattack.[68] With Lang's men pressuring York's 1st Battalion on three sides, they withdrew west off Longstop Hill to the small town of Chassart Teffaha, having suffered 356 casualties over the past three days.[69]

The aborted attempt to take Longstop Hill ended any Allied hopes to seize Tunis before the New Year. More importantly, the fight should have ended the hopeless command and control practices pursued by the Allies. Coordination between regimental- and battalion-sized elements of three nationalities simply did not happen quickly, clearly, or effectively, for a whole host of reasons. Yet, the Allies would continue to throw units into battle ad hoc separated from their division and regimental headquarters until after the Kasserine Pass battles. Given the circumstances, York—only twenty-six years old—showcased amazing combat leadership and was fortunate not to surrender his entire command. For an entire day, only his infantrymen held Longstop Hill, with no heavy weapons or artillery. Lang credited both the British and the Americans for "fighting stubbornly and obstinately," but he had outmaneuvered the Allied light infantrymen thoroughly, making their positions untenable.[70] York's 1st Battalion had done what it could under the abysmal circumstances.

The Allies decided to regroup because of the high losses of equipment and men, lack of air superiority, and poor weather; they would adopt a more deliberate approach to expel the Axis from Tunisia. Bad rains throughout late December turned the clay into a sucking mud, and cross-country movement proved impossible. Furthermore, the air advantage enjoyed by the Axis proved decisive throughout December. The Allied airfields west of the Atlas Mountains suffered more days of inhospitable weather, while the Axis airfields east of the mountains maximized the clear days to produce on-call missions to interdict infantry, artillery, or logistic concentrations.[71] Allied air units did not have this flexibility and flew more structured missions against fixed targets such as airfields and supply depots.[72] A seven-week lull developed on the Tunisian battlefields, and serious action would not begin until 30 January 1943. The 1st Armored Division reequipped its light armor battalions during this time with M5 light tanks and filled shortages of M3 Grants with M4 Shermans. The infantry divisions held defensive positions in western Tunisia and experienced minor combat actions.[73]

Evaluating Armored Warfare Data

Too often, scholars make sweeping generalizations about the U.S. Army's capabilities and effectiveness for all of World War II instead of recognizing the cyclical nature of technology and units. Furthermore, the earlier MTO campaigns began a year and half prior to the bulk of the U.S. Army entering western Europe. This section describes the conditions of armored warfare on the battlefields of North Africa and Sicily for several reasons. Both campaigns' conditions and topography varied notably from later campaigns. Finally, it dispels the notion that better German armament alone dictated victory or defeat when facing the first U.S. divisions to enter combat.

The data in the Effect of Firing First chart highlights the advantages of numbers and firing first in a tactical action. The numbers used for these conclusions stem from the results of eighty-three different offensive tank actions throughout the European Theater of Operations excluding the eastern front. These eighty-three actions typically involved company-, battalion-, and brigade-sized units. They are quite representative of the type of actions experienced by the battalion leadership of the U.S. Army in North Africa and Sicily. Included in these numbers are three engagements that occurred in Tunisia in April and May 1943. The information bears out two primary conclusions regarding armored warfare in World War II: the combatant with numerical superiority and the advantage of firing first was almost always the winner, and the combatant with numerical inferiority in the ratio of two to three or three to four who fired first had an even chance of success.

The most telling component of the bar chart is the twenty wins enjoyed by the combatant that fired first in a ratio of one to one. This point more than any other reveals the advantages of firing first in tank warfare. In these engagements, Allied tanks engaged German tanks successfully at an average range of 750 yards, while German tank crews enjoyed an average range of 1,290 yards in successful shots. Data on kills per rounds fired are unavailable in connection with this study, but the advantage of firing first probably allowed for more effective first shots. Attacks that ended in mission accomplishment sustained loss rates of about 15 percent, while failed attacks sustained an average loss rate of

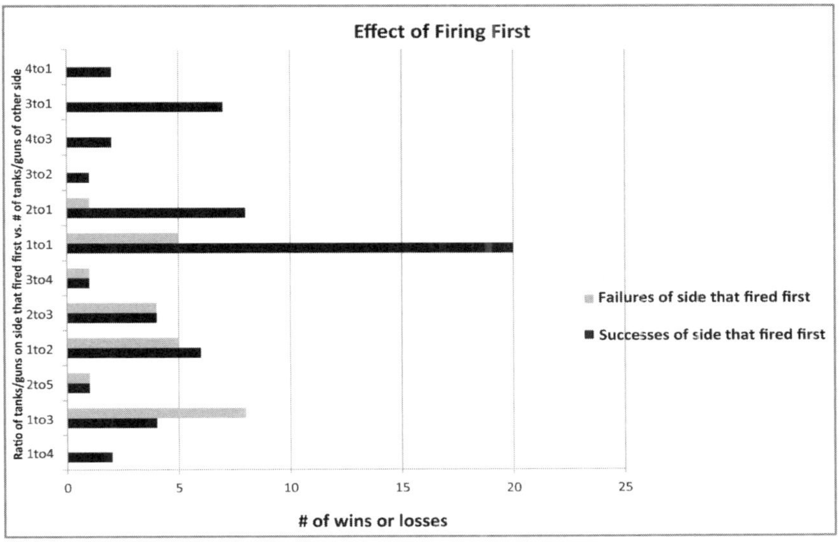

Effect of Firing First. Data taken from the Department of the Scientific Adviser to the Army Council, Military Operational Research Report, Report #33, "Tank Battle Analysis," November 1946, 17–18, MHI.

65 percent. In the eighty-three battles used for this study, the Allied forces were attacking in 70 percent of the actions and usually had numerical superiority. This information would seem to indicate that Allied tanks needed to possess a 30 percent advantage in numerical superiority to succeed against German tanks. This percentage seems a little exaggerated in light of the fact that the Allies had to attack more often and usually possessed more vehicles than their adversaries.[74] The Axis units in North Africa or Sicily did not possess large numbers of advanced tanks, such as Tiger Is and Panthers, that could completely outclass the Sherman or Grant tanks. Within North Africa, Allied forces did not usually have superior numbers of tanks against German tanks until after the battles of Kasserine Pass. The failed race for Tunis as well as the catastrophic destruction of two armor battalions on the 14th and 15th of February reveal how this paucity of armor failed to stop superior numbers of tanks.

Allied tank casualties are another relevant statistic in the importance of armor in North Africa. Eighty-five percent of all Allied tank kills came from gunfire, which includes tanks and antitank guns. Gunfire accounted for 54 percent of Allied tank casualties for the entire ETO, which is 31 percent less than for the North African theater. Why is this important to this discussion? The desert terrain of Tunisia facilitated the ability of tanks to identify and shoot at long distances. While hollow-charge weapons like the Panzerfaust had not yet entered service, their limited range would have had a meager effect on Allied tanks, as did the American bazooka in North Africa. The Axis forces shrewdly used mines to cover their withdrawals through the mountain passes and to protect their infantry strongpoints on key terrain features. Essentially, mines and gunfire accounted for over 96 percent of the losses in North Africa. This point refutes two oft-cited shortcomings of units in the armored and infantry divisions: poor infantry-armor coordination and the dominance of Axis airpower. Most battalion leadership well understood the importance of infantry-armor coordination and frequently commented on it in their lessons-learned documents.[75]

Infantry-armor coordination becomes essential in compartmentalized terrain such as the *bocage* (fields separated by thick hedgerows and sunken lanes) of Normandy or urban centers like Aachen. The terrain of Tunisia did not lend itself to the employment of infantry units except when assaulting strongpoints, conducting night attacks, or occupying chokepoints. The Axis typically located these strongpoints in terrain impassable for tanks; thus, infantry supported by indirect fire usually took positions of this type. Armored attacks in the open dessert such as Sidi Bou Zid or an engagement area defense at Chouigui Pass prevented infantry from playing a substantial role, especially since their only effective armament against tanks was the bazooka, a weapon of limited range and penetration power. The Axis dominance of the skies until late April 1943 did not wreak havoc directly on armored formations, but it did disrupt logistics and provide superior aerial observation. Allied air-ground coordination needed substantial improvement and training, but even if Allied forces had managed to train and develop a sound close air support doctrine prior to the invasion of Tunisia, the destructive effect on the tanks of the Axis would have been minimal. It would have offered advantages in the destruction of infantry strongpoints, truck columns, or antitank gun positions as well as offered badly needed intelligence, but better coordination would not have stemmed the tide of German tanks in the attack at Sidi Bou Zid or against the Blade Force. Waters's Company A endured ten strafing runs in one day, totally exposed to the German planes, and suffered only one casualty.[76]

The next chart compares the average ranges at which Axis tanks destroyed Allied tanks throughout the ETO, excluding the eastern front. The North African front boasts an average range of almost 900

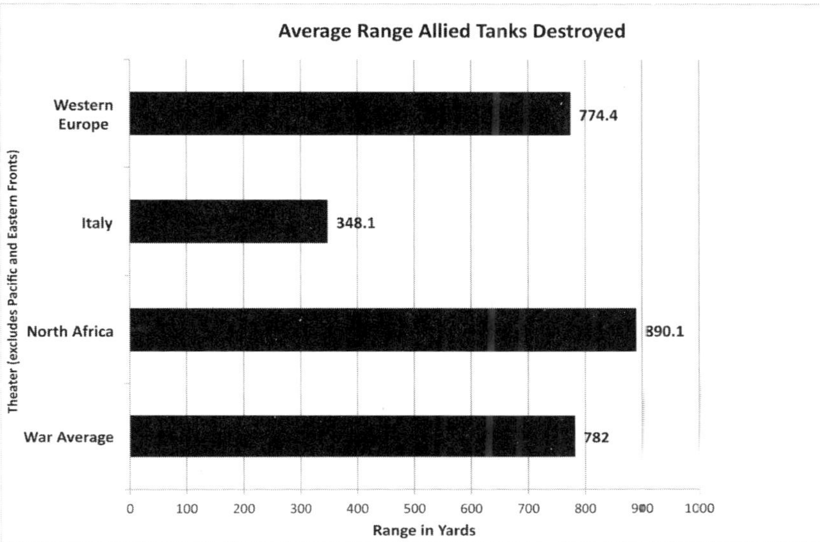

Average Range Allied Tanks Destroyed. Alvin D. Coox and L. Van Loan Naisawald, "Survey of Allied Tank Casualties in World War II," Technical Memorandum, 31 March 1951, 7, MHI.

yards, which is over 100 yards longer than the theater average. The tank's mobility in the open spaces of Tunisia permitted rapid fire and maneuver among armored units. Tanks could often engage or close with the enemy rapidly, without waiting for artillery or air support. The urban, wooded, and mountainous terrain of western Europe and Italy prompted the integration of tank destroyers, antitank guns, assault guns, and hollow-charge weapons in addition to tanks to reduce the average destruction range for Allied tanks. In Sicily, infantry battalion leadership typically employed armor in platoon and sometimes company-sized elements. Outside of the Gela plain in southeastern Sicily, the restricted road network and mountains did not permit any serious movement and maneuver. As a result, very few tank-on-tank engagements occurred. These longer ranges in North Africa demanded that armored formations seize dominating terrain features such as intervisibility (IV) lines or hilltops to overwatch the movement of other friendly forces. Likewise, German armored units used the same fire and maneuver tactics to control passes and plains in the desert. Movement in the desert makes camouflage and concealment quite difficult, adding more importance to these dominant terrain features.

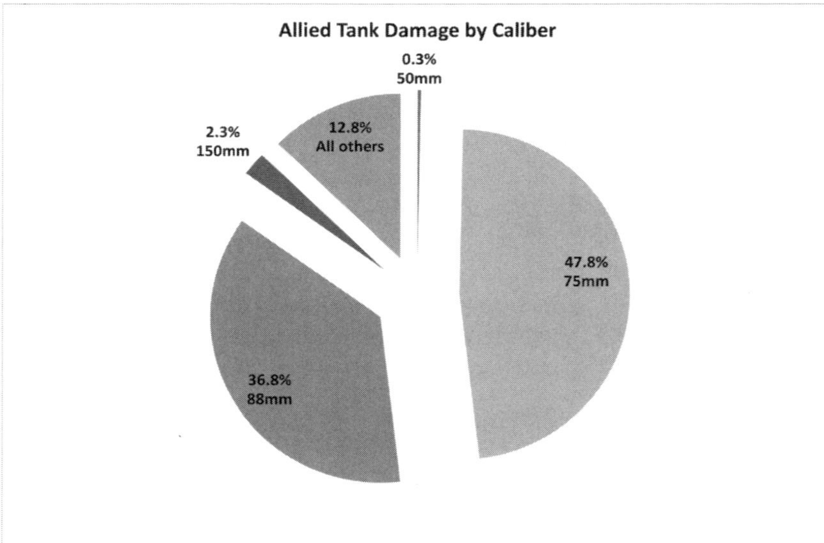

Allied Tank Damage by Caliber. Alvin D. Coox and L. Van Loan Naisawald, "Survey of Allied Tank Casualties in World War II," Technical Memorandum, 31 March 1951, 26–28, MHI.

> The German armoured forces often attacked British unarmoured troops if they found them insufficiently protected by artillery and antitank guns, but they always avoided closing with our tanks in a running fight. When meeting British tanks in strength they preferred to take up a position which was well protected by artillery fire and with antitank guns on the flanks, and used the superior gunfire from stationary tanks to shoot at the British tanks at long range.[77]

The data on range in this study are subjective in that they are based on after action reviews and thirdhand summaries. Given this factor, the average range in North Africa might have been even farther since distance is typically harder to estimate in the desert than in other terrain.

The pie chart above gives approximate percentages of the calibers of guns that destroyed Allied tanks throughout the entire war in the ETO, excluding the eastern front. Two calibers, 75mm and 88mm, accounted for about 85 percent of all Allied tank destruction. The

50mm seems underrepresented given its prolific use as both an antitank gun and main gun armament for the Mark III. Smaller calibers such as the 20mm and 37mm as well as much larger artillery guns make up the remainder of the percentage. This information points to two conclusions regarding armored warfare in North Africa and Sicily: indirect fire had negligible results in the destruction of tanks, and Axis airpower failed to influence armor formations.

> Commanders all agree that the effect of artillery fire on tanks is very slight. Direct hits are the exception and near-misses apparently have but little effect, other than some blast effect, on crews. Tank CO's [commanders], when riding with their heads and shoulders out of the turrets—which is normal—are sometimes wounded by shell fragments and jarred up considerably by the concussion, but unless actually hit, are able to proceed with little ill effect.[78]

While both German and Allied tankers concur about indirect fire, their views on the effectiveness of airpower are in stark contrast. As discussed before, the light Stuart tanks shrugged off Axis airpower even when German aircraft dominated the skies over Tunisia. Allied tank losses indicate almost no tanks lost to enemy aircraft despite the air superiority of the Axis, which they would finally relinquish and never get back. In an interview conducted on 24 May 1945 at the Seventh Army Interrogation Center, German Generals Heinz Guderian, Leo Geyr von Schweppenburg, Josef "Sepp" Dietrich, and Paul Hausser were asked what caused their tank losses. While Guderian failed to mention Allied air, Dietrich listed enemy air as the cause for 10 percent of his losses, Hausser listed a percentage of 20 lost to airpower, and von Schweppenburg said, "Air attacks are very effective and most feared by tank crews."[79] The intent of this assessment is not to compare aircraft capabilities and armament, but merely to point out the contributions that Allied air forces could have made in North Africa and the HUSKY landings if they had possessed air parity, superiority, or temporary superiority.[80] The 10–20 percent tank kills might have been appreciably different in the race for Tunis in late November and early December when the Axis forces moved concentrated formations rapidly from the ports to central Tunisia. Furthermore, a coordinated close air support could have provided critical firepower to outgunned rifle battalions in Sicily during counterattacks immediately after the landings. Airpower was most effective in the complete disruption of logistics and movement of troops. British air dominance had robbed German ability to maneuver, as the Panzerarmee's last offensive against the British 8th Army at Alam Halfa in August 1942 displayed.[81] It led Erwin Rommel to declare that "the unbroken heavy attacks by the RAF, which practically dominated the air space and literally nailed my army to the ground," prevented the Axis from grasping any operational or tactical opportunities.[82] In Tunisia, Axis forces could still operate tactically without significant persecution from Allied air forces. In Sicily, Axis forces enjoyed the same luxury until halfway through the invasion of the island.

Major General Ernest N. Harmon, commander of the 1st Armored Division after Orlando Ward, listed the priorities of the ideal tank: first, gun power; second, maneuverability; third, armor protection.[83] In reality, the U.S. armored force placed most stock in mobility and mass production to deploy as many tanks as possible against the Axis powers and encourage the use of the tank in exploitation breakouts. This type of thinking delayed the debut of the superior Pershing tank to almost the end of the war.[84] The Sherman shouldered the burden of the U.S. armored force from January 1943 until the end of the war, and it changed very little. The most appreciable change made to Shermans occurred when a 76mm high-velocity cannon replaced the shorter 75mm cannon in some models beginning in January 1944. The armored protection increased by only three-quarters of an inch. The M4A3E2 assault tank model, nicknamed Jumbo, which outweighed its sister models by at least 6 tons in added protection, was the only significant up-armored Sherman. Out of 49,234 Shermans produced, Jumbos accounted for only 254.[85]

The next bar chart derives from a study of 107 Sherman tanks evaluated after they were immobilized on the battlefields of Europe or North Africa. Both the German 88mm and 75mm guns had the capability to penetrate up to 5 inches of steel, which made all the Shermans except the Jumbo vulnerable to frontal shots. These two calibers also accounted for almost 85 percent of all tank casualties, but only 30 percent of the Shermans listed received hits on the front of their armor. Hits on the side and rear account for 70 percent of the total from this study. Admittedly, 107 is a small sample of incidents given the volume of tank destruction, but the percentage difference suggests that German tank gunners preferred the more vulnerable areas of Allied tanks and that tactics mattered considerably in the terrain of North Africa. Tanks did not just slug it out, toe to toe, but units exhibited fire and maneuver to gain advantage even when one side had a discernible tech-

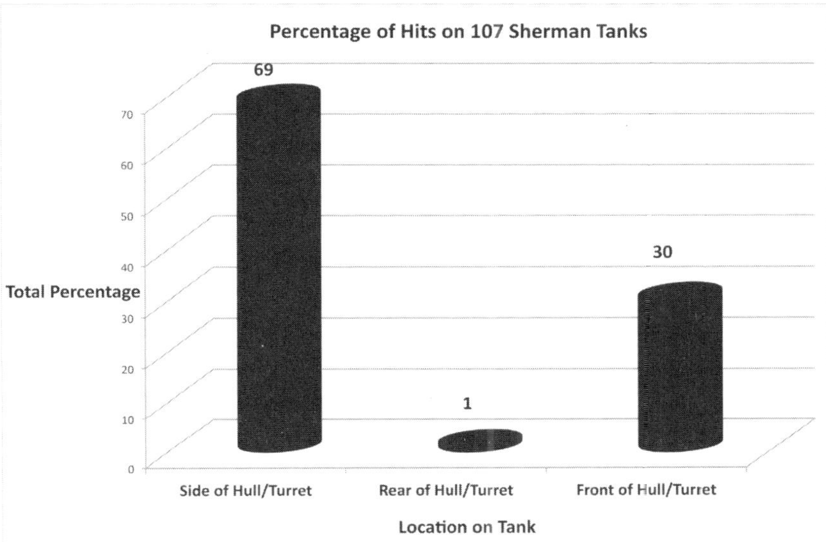

Percentage of Hits by Location of Impact on 107 Sherman Tanks. Alvin D. Coox and L. Van Loan Naisawald, "Survey of Allied Tank Casualties in World War II," Technical Memorandum, 31 March 1951, 22, MHI.

nical advantage. In the battle of Sidi Bou Zid in February 1943, no German ordnance penetrated the frontal armor of the Sherman tanks, but the German tankers focused on the sides and rear of the American tanks. Disparity in armor would not by itself decide any future battles in Tunisia or Sicily with the entrance of the Sherman tank. The Allies had selected the Sherman as their tank for the vast majority of World War II regardless of its weaknesses when compared to advanced models of German tanks. Technological determinism would not override the contributions of battalion leadership in North Africa or Sicily.

Postwar tank scrutiny by battalion leaders confirmed the U.S. faith in maneuverability even after they witnessed heavily armored German tanks as well as the ubiquitous antitank guns defeat American armor in Italy and western Europe. In June 1945, the Army Ground Force Equipment Review Board concluded that the U.S. Army should develop five different tank models and an armored personnel carrier.[86] A light tank—up to 25 tons with a 76mm gun—for reconnaissance and security, a medium tank weighing up to 45 tons with a 3-inch gun for exploitation and pursuit, and a heavy tank with a 75-ton weight and a 90mm gun would comprise the primary leviathans for the armored force.[87] In addition, the AGF Board recommended the development of a 150-ton "super-tank," an armored personnel carrier (APC), and an increased emphasis on amphibian tanks.[88] The AGF Board's report, known as the Cook Report after the board's leader, Major General Gilbert R. Cook, was reviewed by battalion, regimental, and division commanders on 4–6 December 1945 from an armored warfare perspective. Former commanders from North Africa such as Major Generals I. D. White, Orlando Ward, and Lunsford Oliver and Colonels Peter Hains, Louis Hightower, and William Kern participated in the hearings. Of particular note, the board relied heavily on Hightower's testimony.[89] Their review undermined several Cook Report recommendations and revealed the review board's grasp of combined arms warfare.[90]

Prewar armor doctrine and tank development attempted to encourage the U.S. Army to avoid enemy formations in favor of deep penetrations. Both the Cook Report and several older generals on the review board tried to forward this legacy. Commanders, like Hightower, balanced these claims with their experience. They knew U.S. doctrine, but when they encountered heavy enemy formations on the battlefield, U.S. units had to employ tanks against them. This testimony convinced the board members that firepower and mobility should carry equal weight in the design of future tanks and APCs, and armored protection should be a distant third.[91] Although the officers recognized the M3 and M4 inferiority in tank versus tank engagements, they all concurred that they would rather have the mobility over the increased protection of German tanks. Moreover, some officers believed the inferiority was an overall plus. The tank inferiority promoted U.S. Army reliance on combat teams—also known as combat commands—and artillery and close air support (mostly P-47s) integration.[92] Their postwar thought captures the agile thinking developed during war by combat leaders, particularly the battalion leaders. Their testimony reveals their capacity to not stay wedded to doctrine or individual branch initiatives. In another example, the Cook Report and a couple older generals believed that the U.S. Army should not issue maps below the division level. Battalion leaders vehemently refuted this assertion and convinced the board to overturn the Cook recommendation.[93] In order to carry out their commanders' intent, former battalion leaders required the appropriate maps and graphics. In this frank committee environment, armor and infantry battalion leaders such as Hightower and Kern revealed not only what they had learned in battle, but how they thought about war.

Given the severe handicaps stacked against the Allies, it came as no surprise that they could not break the reinforced Axis forces in Tunisia. Even though battalion commanders routinely demonstrated their ability to execute their superiors' orders, it was not enough. Their seasoned maturity and prior training enabled them to synchronize combined arms operations. Above the battalion level, regimental and division commanders were often impotent to affect the battlefield. U.S. battalions were committed without their higher headquarters, inadequate indirect fire, air support, or intelligence. Too often, battalion commanders found themselves trying to work with allies in awkward command structures. They could not rely on trained replacement soldiers or seasoned junior officers without additional training. Finally, they did not possess a superior technological or logistical edge. When the Axis attempted to seize the initiative in February 1943, these handicaps persisted.

5

A Long February in the Tunisian Passes

After the failed race for Tunis, Eisenhower developed Operation SATIN, a plan designed to advance methodically through central Tunisia in order cut off Arnim's Fifth Army from Rommel's Afrika Corps. SATIN was canceled by the rushed thrust to take Tunis. The Allies then established defensive positions centered on the four major passes along the Y-shaped Eastern and Western Dorsals that run in a north/south direction south of Tunis. The Fifth Panzer Army seized the passes from poorly equipped Frenchmen on 30–31 January and defended them from small U.S. counterattacks that never had enough combat power to succeed.[1] These Axis movements preceded Operations FRUEHLINGSWIND (Spring Wind) and MORGENLUFT (Morning Air), conceived as two thrusts to wreak havoc in the Allied rear. Arnim would lead the Spring Wind attack through the Faid Pass with the 10th and 21st Panzer Divisions while Rommel would advance with a division-sized unit of nine German and seven Italian battalions. With no army group yet in existence, both German commanders had to coordinate with each other throughout the operation.[2]

Despite expecting an attack, the Allied command got caught flat-footed when Arnim unleashed his units toward Sidi Bou Zid. The Axis gamble resembled the later offensive known as the Battle of the Bulge in that it achieved operational surprise and was the last gasp of a desperate enemy to cut off the Allied armies at their base. Like the 1944 offensive, it failed and did not possess enough combat power or coherent coordination to accomplish its stated goals.[3] Two biases influenced the intelligence picture of the theater: an overreliance on Ultra by Eisenhower's intelligence officer, British General Eric Mockler-Ferryman, and an obsession that the German attack would come at the Pichon and Fondouk Passes. Subordinate units and staffs, including the II Corps G2 and tactical commanders in the area, believed that a German attack would come through Faid.[4] This assumption prompted the II Corps commander, Major General Lloyd Fredendall, to heed his G2's advice and to take action.

Without ever seeing the ground, Fredendall ordered the positions of specific units that were to occupy on Djebel Lessouda and Djebel Ksaira, which were 10 miles apart and too far apart to be mutually supporting positions.[5] The II Corps order mandated infantry foot patrols and trip wires to help collect reconnaissance as well as specific locations for 1st Armored Division units, essentially robbing Ward of any autonomy as a division commander.[6] The division selected Lieutenant Colonel John Waters to organize this impossible defense since he was the executive officer of the 1st Armored Regiment and considered an extra lieutenant colonel. Ward vented his frustration to Waters about the specific orders issued by Fredendall: "I have no command."[7] Waters had three days to figure out how to defend 10 miles with fifteen tanks, four tank destroyers, and two battalions of infantry. He came to the conclusion that he did not have enough combat power to coordinate a potent defense. The night before Arnim attacked, Waters asked General Raymond McQuillin, the Combat Command A commander and a man woefully out of his depth,[8] at a briefing: "Suppose tomorrow morning I wake up and I find that I'm being attacked by an Armored Division coming through the Faid Pass. What do I do about this?"[9]

Stranded on Lessouda with just a captain, a driver, and a half-track, Waters had evaluated the situation correctly. His command was doomed should the Germans attack in strength. German General Heinz Ziegler, commander of FRUEHLINSWIND, launched his attack against the thin line of Americans at 0630; he caught the tank company unprepared (the commander was still sleeping) and destroyed all fifteen tanks and isolated the infantrymen dug in upon the hilltops. Waters's early warning patrol had occupied the wrong position, and he had not checked it. All he could do was watch the German tanks surround Djebel Lessouda and send updates via his command post half-track to his regimental commander, Colonel Peter Hains. The Germans captured Waters by 1600 that day, and he spent the rest of the war as a POW.[10] His training and combat experience did not fail him. He knew that the positions could not be defended, but poor battle command by an incompetent corps commander and a division commander who did not challenge Fredendall's judgment had set the stage for disaster. Now, a division stretched over 60 miles had to fend off a concentrated attack by two seasoned divisions while trying to rescue two infantry battalions.

Marooned Infantrymen: The 168th Committed by the II Corps

The 168th Regiment from the 34th Infantry Division provided two rifle battalions to establish the defensive positions on Djebels Lessouda and Ksaira under Waters's command. Prior to this assignment, the regiment had served as the II Corps manpower pool for random assignments. By mid-January, the regimental commander, Colonel Thomas Drake, began a 1,500-mile journey to consolidate his unit. After receiving a mission directly from the II Corps commander, Colonel Drake spent the next week locating his rifle battalions for the upcoming attack on Sened Station.[11] He finally consolidated the majority of the unit and led a successful, but costly, attack to seize Sened Station two weeks prior to the Kasserine Pass battles.[12] In this attack, his 1st Battalion commander, Lieutenant Colonel John Petty, was killed, and his 2nd Battalion commander, Lieutenant Colonel Dewey Bauer, was wounded.[13] The regiment consolidated after the Sened action and received its fateful mission to defend forward the Djebels surrounding Faid Pass.

On the 7th of February and under the cover of darkness, the 2nd Battalion, now commanded by Lieutenant Colonel Robert Moore, secured DJ Lessouda, and Lieutenant Colonel John Van Vliet Jr.'s 3rd Battalion occupied DJ Ksaira. The German attack on 14 February had completely isolated both battalions from any other American force. Both 1st Armored Division attempts to liberate the rifle battalions had failed. Marooned on these mountains with no effective communications outside of the regiment, the battalions had few options. On the night of 15 February, Lieutenant Colonel Moore—in command of the battalion for only a week—issued his companies a succinct operations order and led his battalion through German lines under the cover of darkness.[14] Despite running into numerous German units, Moore's efforts resulted in half of his battalion reaching friendly lines. Van Vliet's 3rd Battalion did not enjoy the same fate.[15] Almost 30 miles from friendly lines as opposed to Moore's 15 miles, the 3rd Battalion had a more difficult journey. During darkness, Van Vliet deftly maneuvered his battalion through the German positions.[16] However, as the sun came up on the 17th, over 90 percent of the unit, including Van Vliet, surrendered to elements of the 21st Panzer Division that surrounded the exposed infantrymen.[17] Poorly positioned by II Corps and eventually surrounded by enemy forces, these two battalions suffered from unsound leadership at the operational level. Yet, given these circumstances, both commanders had exhibited courage on the mountains of Faid Pass by keeping their units together and leading them through German lines in open terrain. Similarly, two armored battalions were committed in the same piecemeal manner in attempts to stop the German attack and rescue Moore's and Van Vliet's units. Lieutenant Colonel Hightower's 3rd Battalion, 1st Armored Regiment, positioned about 10 miles to the east, received the mission to stop the German advance. After this failure, Lieutenant Colonel Jim Alger's 2nd Battalion followed up with another failed counterattack. An analysis of both of these armored attacks follows in the next two sections.

Hightower, Leadership Traits, and Defeat at Faid Pass

Lieutenant Colonel Louis V. Hightower jumped off the front slope of his Sherman and hit the ground with a thud. He glanced back over his shoulder to see his tank, which had just scored hits on four German panzers, go up in flames. After the vehicle took numerous enemy rounds to the track, engine compartment, and turret, Hightower had ordered an evacuation of his wrecked tank and escaped on foot west across the Tunisian desert with the rest of his crew. As he picked his way through wadis and past boulders, he knew that his 3rd Battalion, 1st Armored Regiment, had not done well.[18] While he could not have known that his battalion suffered the loss of forty of its forty-seven tanks, he could see the charred hulks littering the landscape, trailing high, dark clouds of smoke fueled by secondary explosions of ammunition and fuel. Tanks from the 10th and 21st Panzer Divisions destroyed his battalion in less than an hour. Hightower counted himself fortunate to have escaped with his life and not ended up a prisoner of war.[19]

Four months later, Hightower sat down to a typewriter in his dimly lit command post on 9 June 1943 and tapped the keys to produce a lessons-learned document for the 1st Armored Division commander about his combat experiences in North Africa.[20] Now the commander of the 1st Armored Regiment, which comprised the core unit of Combat Command A,[21] he must have reflected on the actions at Faid Pass and wondered how to capture the lessons. *Did I fail as a leader? Did my commander fail me? Were my tanks inferior? How bad are our armored tactics?*[22] These must have been some of the questions that he pondered while crafting this document. Whenever a commander loses a fight in combat, he replays the events over and over, looking for those pivotal decisions that he wished he could change. A detailed look at this document shows a battle-hardened leader who objectively analyzed the failures and successes of his experiences. Nowhere in his three-page document did Hightower deride the leadership of his subordinates nor the training of his soldiers. His focus lies with criticism on force structure, superior command relationships, air support, and suggested equipment changes. But what about Hightower? What role did his leadership or grasp of armor doctrine play in the Faid Pass fight?[23]

Lieutenant Colonel Hightower represented the typical midlevel officer of the 1st Armored Division in North Africa, and his grasp of leadership and mechanized warfare should reveal the reasons why the 1st Armored Division failed at Kasserine, but fought successfully against the Axis powers prior to and after the Kasserine battles until the Allies expelled the Germans and Italians from the shores of North Africa. At the time of the Faid Pass fight, Hightower was a thirty-three-year-old West Point graduate from Texas. A midlevel officer typically commands battalions or serves as the executive officer or operations officer for battalions and regiments that contain from a few hundred men up to a few thousand soldiers and hundreds of vehicles and weapons. In order for these sizable units to coordinate their efforts in the melee of battle, their commanders must display sound leadership in preparation for and during combat as well as maintain a firm grasp on how to employ and adjust doctrine based on battlefield conditions. Lieutenant Colonel Louis Hightower serves as a case study to understand the complexities of leadership, doctrine, and personal characteristics of combat leaders. So, what led to the destruction of Hightower's battalion at Faid Pass? Hightower displayed outstanding bravery and sound leadership as well as an impressive grasp of U.S. Army armored doctrine during the battle. It was the failure of adequate battlefield intelligence that prevented him from the execution of a more successful operation.

The midlevel officers of the 1st Armored Division displayed common traits that allowed them to be effective combat leaders in the battles of North Africa. Effective leadership among U.S. Army officers encapsulates a diverse array of intangibles. Leadership designates or encompasses an officer's decision-making abilities, bravery under fire, and communication skills. The vast majority of the battalion leaders exhibited effective leadership traits in combat. These leadership values and the comprehension of doctrine acted as guiding principles for officers such as Hightower.

The use of doctrine includes an officer's understanding of written field manuals that address tactics, techniques, and procedures, technical specifications and limits of equipment such as a radio or an M4 Sherman tank, a grasp of the logistical requirements to maintain a unit's

American Tank Desert Camouflage in Tunisia. This is a photograph of one of the tanks from Hightower's battalion. Notice the ad hoc camouflage. "Desert camouflage is a little light colored for Northern Tunisia, N. Africa, 1943," Photo 7-9, Robert E. Van Zant Collection, Photographs, General George Patton Museum, Fort Knox, Kentucky.

readiness, and the knowledge to train these tactical, logistical, and technical aspects prior to combat. Each officer's ability to lead and execute doctrine differed.[24] The failure to adapt to combat conditions usually increased one's chances for quick removal, capture, or even death.[25] Hightower epitomized the effectiveness of battalion leadership of the 1st Armored Division officer corps in Tunisia.

Lieutenant Colonel Hightower received the unenviable task of securing the Faid Pass after advancing elements of the 10th Panzer Division destroyed Company G, 3rd Battalion, 1st Armored Regiment, at about 0600 on the morning of 14 February 1943. Unfortunately for Hightower, he could have used this company later. In addition to this disturbing event, two infantry battalions located on mountains securing this pass were in danger of being completely surrounded and had already been overwhelmed by superior German firepower and numbers. Hightower moved his battalion, consisting of two tank companies and part of A Company, 701st Tank Destroyer Company (75mm guns mounted on half-tracks), east to hold the pass. He parked his tank, named *Texas*, on top of a small hill to observe the enemy movement around the Faid Pass. From this vantage point, he observed at least one hundred German vehicles advancing in clouds of dust from the east. The time was about 0930, and Hightower reported to his superior, Brigadier General Raymond McQuillin, that the Germans decisively outmatched his force and that he could only fight a delaying action. Hightower's battalion engaged the converging forces of the 10th and 21st Panzer Divisions northeast and west of a town called Sidi Bou Zid. Unknown to Hightower, this town served as a primary objective of the attacking Germans, and his tanks received the brunt of their attack. The relentless onslaught of German panzers and antitank guns slowly bled Hightower's battalion, but the deft maneuvers executed by the 3rd Battalion, 1st Armored Regiment, allowed the rest of the 1st Armored Regiment to escape Sidi Bou Zid before the Germans encircled the town. Hightower held his ground against the 10th Panzer Division's advance from the east, but he was overwhelmed when the 21st Panzer Division's lead elements attacked from the south.[26]

The radio operator of *Texas*, Sergeant Clarence Coley, could not contact any of the remaining tanks of 3rd Battalion, so Hightower recognized the desperate situation and maneuvered his lone tank into position to engage the seven enemy tanks advancing from the south. Hightower ordered the driver to stop and popped up out of the tank commander's hatch with his binoculars to spot his gunner's shots. The tank crew feverishly worked to destroy the panzers as they advanced toward this position. Soaked in sweat, the loader shoved round after round of 75mm ammunition into the breach of the gun every three to four seconds. Hightower and his gunner, Corporal Bayer, worked together to adjust their shots. *You shot over his turret! You got him with that one!*[27] As the ammunition racks dwindled, Sergeant Coley scrambled to dig additional ammunition out of the hull of the tank while German rounds made thundering shocks against the skin of the Sherman. The Germans kept getting closer, and the anxiety grew among the crew cramped in their steel box filled with the acrid smell of spent shell casings, fuel, and sweat. Then over the din of the rumbling engine, the loader, Private Agee, yelled that a round got stuck in the gun. Defenseless, Hightower ordered the tank to move to a more protected location, but before the vehicle lurched anywhere, a German shell penetrated *Texas*'s turret. The round punctured the gas tank, leaking fuel all over the crewmen and ammunition, and ricocheted through the tank, narrowly missing everyone. Sergeant Coley recalled Hightower saying, as the projectile laid there spinning and sputtering fire, "Let's get the hell

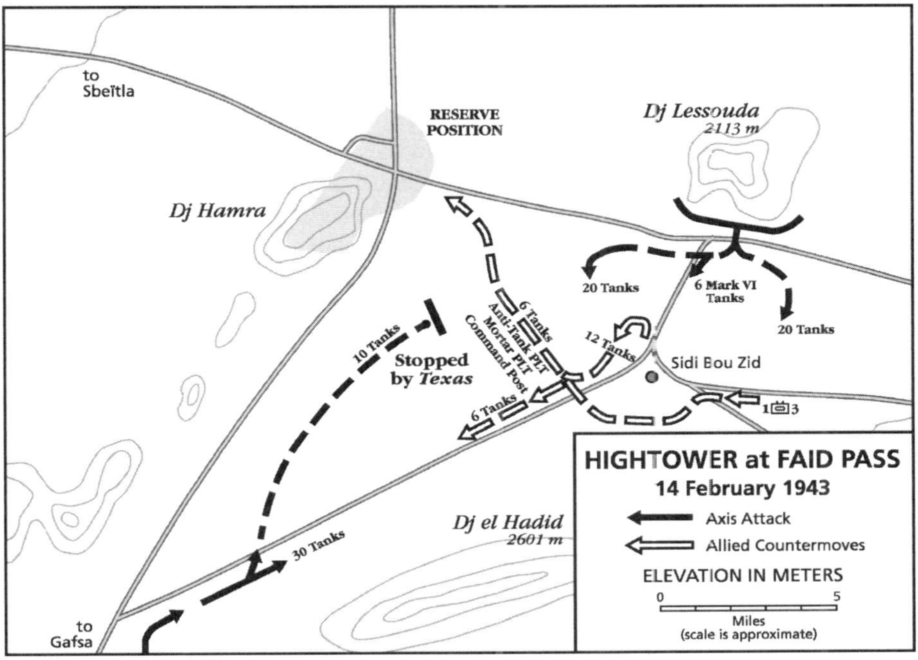

Lt. Col. Hightower Sketch of Faid Pass. Sketch drawn by Lt. Col. Hightower of the last withdrawal from the Faid Pass in which he lost his tank. The fight originated much farther east against elements of the 10th Panzer Division and ended with his last stand against the lead elements of the 21st Panzer Division shown in the center of this sketch. From Operations of 3rd Battalion, 1st Armored Regiment, from January 1, 1943 to February 21, 1943, 30 July 1943, Sketch 3, RG 407, NARA II, College Park, Maryland.

out of here!" As *Texas* erupted into a fireball and the crew sprinted away, the Germans halted their advance to regroup after the one-tank stand orchestrated by Hightower.[28]

The road west of Sidi Bou Zid lay unprotected, and the German units could have attacked unmolested. Hightower contributed to two documents immediately following the action at Faid Pass that presented lessons learned from this engagement. Typically, documents of this kind tend to focus on the negative experiences of a leader in the hope that someone reading the document will not make the same mistakes. Lessons-learned documents normally do not recount glorious feats on the battlefield, and they are generally not rosy recollections of a fight.

The intent of those documents is to highlight flaws and sound practices in doctrine for future engagements. A review of Hightower's comments reveals an officer who possessed a command of fighting in the desert against the Germans.

Remembering the loss of *Texas*, Hightower must have regretted the senseless slaughter of his Company G, detached from his unit prior to the battle of Faid Pass. This unit could have provided much-needed combat power during the battle or significant early warning of the German attack through and around the Faid Pass. Hightower believed that his commander, Brigadier General McQuillin, should have never detached one-third of his unit from 3rd Battalion. The following comment supports this: "In operations requiring an attack by a battalion the battalion commander should be given all of his tank companies in order that he may have a support company, and a reserve. Normally the sector assigned requires two companies deployed in line. A tank reserve should be held out by the battlefield commander under his control. Therefore he should have a sufficiently strong force to enable him to do so."[29] Two more points made by Hightower signal his frustration with his superiors not adhering to doctrine. "We must learn the principle of being strong at the right spot and avoid trying to hold everywhere" reflected his belief that the entire 1st Armored Regiment's frontage of 25 miles at the outset of the Faid Pass fight overextended the unit's assets.[30] Second, the complete lack of guidance and intelligence handed down from Hightower's higher headquarters must have made him remark, "Higher commands must inform lower echelon commanders of the general plan for offense or defense in order to get intelligent cooperation."[31] These remarks indicate that Hightower had a solid grasp on the integration of his command with his higher headquarters, the 1st Armored Regiment. Hightower conveniently failed to explain why he did not demand better information or a reserve from his boss. Perhaps in hindsight he realized he should have more clearly asked for more favorable conditions for his attack at Faid. More importantly, Hightower's astute observations fall in line with the U.S. Army field manuals in publication at the time.[32]

According to two field manuals on armored operations, *FM 17-33* and *17-10*, "the regimental or combat command commander is responsible for coordination of the attack as a whole,"[33] and "although the regimental commander may place much responsibility upon his subordinate commanders, he must still, personally and through his staff, coordinate and direct their actions."[34] Hightower's attack toward the Faid Pass received paltry support and coordination from the 1st

Armored Regiment headquarters. The view of this action through Hightower's eyes gives a clear vision of how doctrine is, ideally, supposed to work on the field of battle.

An in-depth look at doctrine clearly indicates that most leaders considered it sound and effective on the North African battlefields. At the end of the 1st Armored Division's campaign, Major General Harmon concluded that there should be no changes to present tactical doctrines and added, "The only thing that is required is intensive training in the execution of those doctrines and the development of the proper leadership to put them into effect on the battlefield."[35] Of the 160 field manuals in print as of February 1942, the army had developed or updated all of them but eight since 1940. Of those eight dated prior to 1940, only one—*Signal Communication, FM 24-5*—stands out as a serious oversight since the other seven dealt with sketching or 75mm field artillery gunnery.[36] These manuals were distributed down to the regimental level for officers to study. Where no doctrine existed, such as for reconnaissance units, commanders improvised their actions based on their assigned missions as well as the utilization of overarching manuals such as *FM 100-5, Operations*.[37] For example, the 81st and 91st Cavalry Reconnaissance Squadrons executed raids, established observation posts, patrolled both mounted and dismounted, and built defensive positions throughout the North African campaign.[38] The seven Cavalry School studies built from their actions frequently reference *FM 100-5* to validate the principles employed by these units. Tank destroyer doctrine received the only significant criticism from the field.[39] Battalion leadership could rely on their doctrine to visualize the way they would fight their units in battle.

No record exists of the exact radio communications or briefings that Hightower issued to subordinate company commanders prior to and during the battle for Faid Pass. While the battalion lost forty-four tanks, it did tie up the German attack from about 0930 to 1500. The Germans maintained local air superiority throughout the engagement, and none of Hightower's requests for air support ever came to fruition.[40] Throughout the engagement, he maintained a position on the battlefield that offered observation of the enemy and his forces. He attempted to reposition his two tank companies when they sustained heavy casualties, but superior German firepower and enemy air support negated his efforts. His inability to communicate with his tank companies as the battle wore on did not deter him from leading his soldiers up front. Based on Hightower's own account and his radio operator's, his decision to attack seven German tanks on his own attests to his bravery under fire. He displayed a clear head in evacuating his crew after the vehicle sustained a catastrophic hit and not only directed his crew to destroy four German panzers, but led them to safety after the loss of *Texas*.

Based on the analysis thus far, Hightower exhibited the requisite leadership traits that were part of effective combat battalion leadership and adequately grasped the doctrinal shortcomings that affected the action at Faid Pass. Another significant issue regarding this officer is his knowledge of armor doctrine as it applied to soldiers and officers under his command. The Faid Pass fight showcases how a battalion leader was forced to adapt to less than ideal combat conditions. Hightower conducted little or no reconnaissance and did not have access to other assets such as a reconnaissance airplane to inform him of just how extreme the situation had become before he entered the fray. Company I, one of his two tank companies, received extremely accurate fire from German antitank guns, and Hightower was unable to silence these guns or reposition Company I before their tanks ended up as smoldering hulks. Last, Hightower's brave one-on-seven fight separated him from commanding his companies and denied him sight of the whole battlefield. Hightower addressed all these issues in an interview on 1 March 1943, probably after reflection on mistakes he committed or the enemy forced him to commit. Perhaps the most telling of his lessons is: "Teach your commanders to stay out of the fight until they are the last tank or thereabouts. They are too prone to become interested in a personal duel and forget about their control of the units."[41] Of course, no leader performs perfectly under the stress of combat, and mistakes happen based on numerous factors, but Hightower's lessons-learned documents reveal an officer aware of the intricacies of armored doctrine.

Two accounts, one German and one American, compiled shortly after this engagement confirmed the valiant actions of Hightower and his battalion throughout their delaying action at the Faid Pass. Major Helmut Hudel, of the 10th Panzer Division, remarked that "the enemy fought back hard and tenaciously," and the "SHERMANS [of 3rd Battalion] appeared continuously firing from shortest distances, which at times were only five to ten meters."[42] Hudel's perspective on this fight specifically details how the German armored units had to employ their tanks in wide envelopments as opposed to more direct fire and maneuver tactics[43] because they feared the superior firepower and adept deployment of Hightower's M4 Shermans. As the German Mark III, IV, and VI tanks outflanked the smaller force, "engines were strained to their utmost capacity," and "this maneuver succeeded to split up the defensive fire" of Hightower's tanks.[44] Despite the overwhelming victory achieved by the Germans, they halted their attack at Sidi Bou Zid to consolidate their forces and prepare for an American counterattack. The 3rd Battalion's actions denied the initiative to the attacking Germans and secured a short reprieve for the rest of the armored division.

Colonel Paul Robinett provided a coherent and objective description of the 3rd Battalion, 1st Armor Regiment, at Faid Pass. Robinett commanded Combat Command B of the 1st Armored Division in North Africa, which consisted mostly of the 13th Armored Regiment. He would successfully lead the defense west of the Kasserine Pass one week after this fight. Although he defended the passes at Maktar, 80 miles to the north of Faid Pass, while 21st Panzer and 10th Panzer enveloped Hightower and his battalion, Robinett interviewed many of the officers and soldiers who participated, and he had access to all the planning and lessons-learned documents for the 1st Armored Division. Without "the determined action of the Battalion Command, Lieutenant Colonel Louis V. Hightower, there is no question but that the entire Combat Command [A] would have either been killed or captured."[45] As German armored vehicles moved in large formations through Faid Pass, north around Djebel Lessouda, and from the south of Faid Pass, "it was immediately obvious, that due to the overwhelming superiority in numbers and in fire power of the enemy tanks, that the 3rd Battalion would be unable to halt the German advance" west.

"Therefore, the Battalion Commander [Hightower] ordered a withdrawal toward Sidi-Bou-Zid, delaying from position to position."[46] Hightower later remembered:

> We sat there and shot at them, knocking out about five or six while losing three ourselves. Then the twenty-two tanks came over the hill toward us and turned out to be fifty, including at least four of their giant Mark VIs. We drew back to cover as they tried their usual trick of envelopment. We kept pivoting back and shooting at first one German flank and then the other. We kept from being surrounded, but at a heavy cost. Then their dive-bombers caught us, coming over in wave after wave.[47]

All three of these perspectives reveal an overextended and outnumbered tank battalion engaged in a battle with no intelligence about what size forces they encountered west of Faid Pass until both forces began exchanging high explosive rounds. The Germans isolated two infantry battalions guarding the Faid Pass and destroyed every tank of G Company, 3rd Battalion, screening forward of Combat Command A. Hightower led his battalion forward into the unknown. What precipitated this vacuum of battlefield intelligence?

Beginning at the tactical level, Hightower and McQuillin did not develop an aggressive reconnaissance plan that could have provided early warning to identify the movement of two German armored divisions. Lack of manpower and orders from above prevented them from doing this. Fredendall had placed the force under Lieutenant Colonel John Waters in a hopeless position and forbade them from conducting reconnaissance any further east. Despite these limitations, Hightower later commented that "we were completely aware of the build-up of German armor behind Djebel Krechem and the Matleg pass area. We had also predicted, rightly as it turned out, that a major German attack would be launched through the Faid, Rabaou, Kralif, and Maizila Passes. We had planned and discussed all contingencies we could think of the evening before."[48] Even though the Allied commander, General Eisenhower, visited this exact sector of the battlefield only about four hours prior to the German attack, the limited intelligence collected by Hightower's battalion and Waters's units guarding the pass escaped the attention of higher-level commanders. Major Helmut Hudel identified this intelligence gap as a weakness and plainly said so in his conclusion: "American security outposts had not been advanced," which created favorable conditions to stage attacking forces within sight of the east side of the Faid Pass.[49] This massive oversight points to inexperience among the Americans, but the 1st Armored Division occupied a front of over 65 miles. Even with the increased capabilities of modern U.S. Army mechanized divisions, doctrine specifies a 30-mile frontage as appropriate for defensive operations. Even if that frontage is doubled, the 1st Armored Division still exceeded it by 5 miles. Combat Command A had to defend almost 30 miles of this frontage and could not generate enough resources to collect adequate reconnaissance over such an area. Air reconnaissance had to make up the difference.

Returning to Hightower's lessons-learned document: he gave harsh criticism to air reconnaissance for its failure on the 14th of February. "Air observation must keep track of hostile tank moves. It is inexcusable to permit the assembly of over 100 tanks in an area without some warning to our own troops facing them."[50] As discussed before, the poor weather west of the Atlas Mountains prevented the majority of Allied air from the execution of routine or on-call reconnaissance flights. Hightower further indicates that ground troops identified enemy troop movements by dust clouds and sounds at night. When

Combat Command A notified higher headquarters of these observations, no air assets located the probable avenues of approaches or objectives of the 10th Panzer and 21st Panzer divisions. Both Hightower's inadequate efforts on the ground and the dearth of aerial reconnaissance combined in this case to keep the Allied efforts in the dark. Without delving into the complexities of staff coordination at the division and corps level in regard to air support, the fact remains that the end state of visual contact of the enemy through reconnaissance did not happen in a coherent manner for Hightower. Had he received an accurate estimation of the strength of the units advancing on Faid Pass or had early enough warning, then the entire Combat Command A would have withdrawn to consolidate its combat power and string together a more competent defense. No matter how valiantly a unit fights or how well a leader employs sound tactics, a dearth of information can negate these abilities. Also, the equipment on both sides had comparable capabilities, increasing the critical importance of good intelligence.

Lieutenant Colonel Louis Hightower does not fit the image of a green officer unable to lead under fire or prepare his soldiers for battle. The actions of his tank alone dispel the notion of a lumbering peacetime officer unequal to the task of combat. His battalion fought well enough to convince two German divisions to consolidate their gains instead of pushing an attack that probably would have caused more damage. While the 3rd Battalion, 1st Armored Regiment, did not always execute their actions according to doctrine, they understood it and had trained utilizing it. Besides, combat is not a controlled situation, and doctrine must adapt under the stress of war. Hightower has served as a case study in this chapter to reveal the components of leadership, doctrine, and personal characteristics of midlevel officers in the 1st Armored Division. The Hightower study indicates leadership faults above him as well as faulty intelligence collection methods, not a failure of the doctrine or leadership; these factors account for the destruction of the 3rd Battalion at Faid Pass. His personal courage and loyalty to protecting his unit clearly stand out in his use of his own tank against the seven German panzers in order to save his comrades and allow the remainder of his battalion to escape.

At the end of the day on 14 February 1943, the German Fifth Army had destroyed the majority of Hightower's tank battalion, two artillery units, and a reconnaissance battalion. The Germans controlled the Faid Pass and sat poised to continue their attack west. Two U.S. Army infantry battalions remained stranded on hills adjacent to Faid Pass, and their position was untenable as the Germans continued to consolidate their forces. The 1st Armored Division struggled to defend a 60-mile front with fewer units and had to mount an offensive to relieve the two stranded infantry battalions. Major General Lloyd Fredendall issued instructions to General Ward: "As regards action in the Sidi Bou Zid area, concentrate tomorrow on cleaning up situation there and destroying the enemy. Thereafter collect strong mobile forces in Sbeitla area ready for action in any direction."[51] Ward assembled a force to comply with Fredendall's orders as well as to rescue the beleaguered infantryman astride the rocky hilltops of Djebel Ksaira, Garet Hadid, and Djebel Lessouda around Faid Pass. He tapped Lieutenant Colonel James D. Alger, a twenty-nine year-old West Point graduate from Massachusetts, to lead the assault with his fifty-two tanks and a company of tank destroyers. An artillery battalion (minus one battery) and Colonel Stack's 1st Infantry Battalion, 6th Infantry Regiment, trailed Alger's units to provide support as needed. Alger's mission seemed a straightforward attempt to seize the town of Sidi Bou Zid and establish security so both infantry battalions could withdraw from their rocky outposts. It was anything but simple or straightforward.

Alger, Overwhelming Numbers, and the Defeat at Sidi Bou Zid

Alger was representative of the midlevel officers in the 1st Armored Division, and his actions at Sidi Bou Zid confirm that he possessed the requisite leadership qualities and grasp of doctrine of a competent armor officer. Essentially, General Ward employed a flawed plan and did not provide essential support such as air cover or intelligence to set favorable conditions for Alger. Alger employed a potentially effective formation according to armored doctrine and did not falter in leading his men under fire until he was captured. His performance confirms the mettle of the 1st Armored Division officers and the military culture from which he derived these strengths.

Lieutenant Colonel James Alger's battalion served initially as part of the Combat Command B in the Maktar area much farther to the north of Sidi Bou Zid. While in this area, Alger's 2nd Battalion, 1st Armored Regiment, conducted rigorous reconnaissance of the entire Ousseltia Valley and trained for the defense of this valley since most U.S. forces anticipated a German attack through this area. However, very late on 13 February 1943, Alger received the order to move to 1st

Armored Division headquarters in the Sbeitla area. He received these orders almost three days after division issued them. This delay in orders seems like a common miscommunication, but it serves to point out a critical turning point that had unfortunate consequences later.[52]

The battalion could have executed its move to the Sbeitla area on the morning of the 11th. Instead, the battalion arrived at 2300 on 14 February 1943 due to the failure of the division headquarters. The next morning, Alger had to attack with a unit that had spent the previous day under the stress of a long movement with little to no sleep. In addition, the unit had no time to conduct thorough reconnaissance, rehearsals, or preparations for their attack as they had in the Maktar area. Alger lost the most precious resource available to a commander—time.

Compounding his frustrations was the greenness of his unit that had yet to experience ground combat, although they had repeatedly dodged the aerial attacks of the Luftwaffe. Combat units typically perform better in their first engagements if they possess knowledge of the terrain and enemy, and have rehearsed their plan of attack. Furthermore, the 2nd Battalion arrived in a poor posture to conduct an offensive movement. The movement from Maktar to Sbeitla affected their maintenance and logistic readiness. Alger and his officers later remembered, "the rate of march was cut to eight miles per hour after dark. Heavy sand along the route hampered movement south."[53] The last-minute move ensured that the focus would remain on the preparation of vehicles, weapons, and soldiers' eyelids and stomachs instead of the tactical plan.

Commanders must continually coach and mentor their subordinates to develop their leadership skills and execution of doctrine. Alger's diligent preparation in the Maktar area indicated that his immediate superior, recently promoted Brigadier General Paul M. Robinett, had attempted to mentor Alger. Robinett's Combat Command B was the only portion of the 1st Armored Division, besides Lieutenant Colonel Marshall's 3rd Battalion, 6th Infantry Regiment, that had fought ashore at Oran and across Algeria to Tunisia. Robinett's prior experience and exposure to combat probably drove him to teach his subordinates.

> Colonel Alger's 2d Battalion, 1st Armored Regiment, was given the benefit of all we had learned of the enemy, of the terrain, and of precautions to be taken against the Germans' powerful guns. I cautioned him against precipitate action or "rat racing" as it was called in the Division; but I doubt that he really understood the

> power of the enemy's guns. A few platoons of the outfit also engaged enemy targets by indirect fire in the Pichon area, but events to the south soon brought this instruction to an end.[54]

This statement indicates that Robinett grasped his role as a mentor, although he lacked confidence in Alger. This negative reference might imply that Robinett knew that Alger would eventually leave his unit, and maybe he did not view Alger as one of his own. Nonetheless, he made the effort to lead and train the young West Pointer and prepare him to face the Germans. Robinett believed in the superiority of the German veterans until Americans experienced combat for themselves against the Germans. He issued the following comment while he watched Alger's battalion begin its move to Sbeitla: "The outfit was most impressive, but I felt sorry for the men as they moved toward their first battle with the enemy's clever veterans established in positions of their own choosing."[55] Whatever Robinett's opinion of Alger, the time together certainly influenced Alger, and he could have used more planning time quite effectively based on prior mentorship. The second part of Robinett's statement stands out as a dire foreshadowing of the events to come, but also indicates his grasp of the fighting in this terrain. The magnificent observation offered by the wide-open spaces of Tunisia always favored the combatant who occupied advantageous terrain first and defended from it.

Alger arrived at a farmhouse a little after 2300 on the 14th of February to begin planning the operation with Colonel Stack. Around 0400 the next day, the commanders issued the order for the attack. Unfortunately, the order possessed only the basic elements and lacked any significant detail on the enemy situation. Based on Combat Command A's engagements on the 14th, the division intelligence officer guessed that about sixty enemy tanks occupied the Sidi Bou Zid area. Alger had no 1:50,000 maps of the area and only two 1:100,000 maps. While any map is better than none, 1:50,000 maps include detailed information about terrain features that affect cross-country movement, while 1:100,000 maps do not show specific information on features such as wadis and elevations. This factor contributed to the poor distribution of forces during the actual attack, when they encountered unexpected wadis. No detailed data existed of the terrain or enemy between the initial attack position and Sidi Bou Zid, separated by 10 miles of desert. Colonel Hains, the 1st Armored Regiment commander, further compounded this issue by stating that the wadis were passable, although during the actual battle, the tanks had to cross single file across a wadi, which slowed their advance, permitting the Germans to observe them longer. Right before Alger began his assault, Hains and Hightower met him to give some advice on the tactical situation. They warned him of the German practice of concealing antitank guns in "outlying houses" to break up tank formations. The Germans often used Arab garb to disguise their soldiers and gun positions, although Alger had not yet experienced this. For some reason, Hains and Hightower believed that Alger had a four to one advantage in tanks and told him so, but that did not prove to be the case.[56]

From approximately 0600 until 1300, Alger readied his unit for the attack. The units moved into the attack position, also known as the "jump-off" position, and prepared for the attack. This preparation included a two-hour confirmation of the direction of attack, position of units in the formation, distribution of limited maps, and other details. In the middle of this, the Germans effectively bombed a key crossroads with Junker 87s to make it impassible for vehicles. Alger launched the attack at 1300, and his battalion lumbered across the desert to Sidi Bou Zid. A flight of Junkers 87s knocked out a tank destroyer platoon guarding the left flank of Alger's battalion as it attempted to cross a wadi and enter the town of Sadaguia. Immediately following this action, the battalion encountered its first wadi. This gulch proved particularly deep and wide, and only one crossing sufficed as an avenue for the entire battalion to advance. Detailed 1:50,000 scale maps might have forewarned the unit of this perilous obstacle, but no one possessed any details of these wadis. The battalion had to transform from an attack formation into single column. This maneuver decreased the momentum of the attack to an agonizing crawl and provided an opportune moment for the Germans to ascertain the size and location of the attackers. No doubt this vehicular movement churned up an immense amount of the red dust so dreaded by all who survived in the Tunisian desert. After twenty minutes, the 2nd Battalion reestablished its formation, but slowed to traverse a second wadi in their path. At this point, Alger slowed the rate of march not only to cross the ditch, but also to wait for a promised air attack on Sidi Bou Zid, which should have provided key information on the enemy's disposition as well as destroy German positions. The Germans began to disrupt the attack by firing air-burst artillery at the most likely crossing points of this second wadi. Undoubtedly, the Germans had preregistered their artillery in order to cause confusion. Airbursts rain down hot, lethal pieces of shrapnel that force armored vehicles to "button up."[57] Typically, tank commanders and drivers keep their heads and upper torsos exposed to retain situational awareness. The threat of metal tearing into flesh discourages this behavior. Air-burst artillery can also disable radio antennas, vision blocks on the vehicles, and machine guns. Despite this tactic, Alger's D company located and destroyed a German gun position consisting of four 75mm guns and two 47mm guns that had positioned itself to take advantage of the melee caused by the wadi crossing. The tanks climbed out of the second wadi and made for the third and final wadi before the town.[58]

Alger left a tank company in reserve at the second wadi and crossed the third wadi into Sidi Bou Zid still under air-burst artillery fire. By 1530, Alger seized the town after the destruction of multiple machine-gun positions, destroyed two German tanks himself, continued to move east to rescue the infantry, but still had not received any air cover. The attack had achieved one of its two objectives thus far, but the success ended there. Between 1545 and 1555, subordinate units reported enemy tanks both north and south of the town.[59]

Alger must have felt pleased to have done so well with so little preparation. His unit crossed all three wadis and seized Sidi Bou Zid with little difficulty. As he maneuvered to address the issue of enemy tanks, his radio flooded with reports from his units about contact with tanks, losses, and German tanks maneuvering outside of their tank gun range. Alger ordered his battalion executive officer, Major William Emory, to attack to the south with the E Company reserve while the rest of the battalion attacked away from Sidi Bou Zid. Alger headed at full speed in his Sherman to the north of the town to address his left flank when artillery shattered his battalion radio net antenna. He now had only one radio to communicate with subordinate units as well as headquarters, when he really needed two radios. D company lost its commander, Captain Province Winkler, and no one could reach F company to the south of town. Alger struggled to make sense of where his units were and where the German units were located. Within a minute, his tank suffered three hits by antitank guns or tanks, which destroyed his tank's main gun and jammed his turret so his crew was unable to rotate it. Fifteen minutes later, while still crawling defenseless to the north, Alger's tank received two more enemy rounds to the engine compartment that rendered his vehicle immobile. As it still made a juicy target for some German gunner, two more rounds ripped through the turret, killing the radio operator. These last shots convinced Alger of the futility of his situation, and he abandoned the steel coffin to move on foot toward D Company's position. Before he got to them, German tanks and infantry forced Alger and his remaining crew to surrender.[60]

Like Hightower, Alger attempted to recall what happened to his unit after its destruction in Tunisia. Unfortunately for Alger, he did this as a prisoner of war. The Germans collected eighty-one American prisoners throughout Alger's battalion attack in the vicinity of Sidi Bou Zid. All the officers including Alger spent the remainder of the war as members of Oflag 64 (Offizier-Lager, or officer camp) in Schubin, Poland. In an attempt to ascertain the catastrophic events that befell 2nd Battalion, 1st Armored Regiment, Alger coordinated a complex network to record the events as accurately as possible given the austere conditions and limited resources. His efforts coupled with those of his officers resulted in a first-person, chronological summary of the battle. The officers utilized a code in their notes to protect the names of soldiers and units to prevent the extraction of useful information by German intelligence units. They also kept these notes hidden and in separate sections so the entire manuscript never became compromised. After Germany collapsed, Alger retrieved all the sections of their battle chronicle and produced a useful account of the events. While this report holds the most detail and information on the American side, the German perspective offers some insight into the preparedness of Alger and his 2nd Battalion.[61]

The German force contained over 120 tanks, including the new Mark VI Tigers, which Shermans could not penetrate from the front, as well as numerous camouflaged positions of antitank guns. The grim reality overshadowed the woefully inadequate estimate of sixty tanks proffered by intelligence and commanding officers such as Colonel Hains. The 21st and 10th Panzer Divisions under the command of General Ziegler had prepared a careful defense in anticipation of a large American counterattack. The counterattack against Alger's forces achieved overwhelming success because the Germans delayed their movement against what they believed was a very small American unit. They waited to unleash the bulk of their armor until they confirmed that Alger's force was alone. This delay drew Alger's battalion deep into the jaws of a massive double envelopment. The German tanks and antitanks handily destroyed all fifty-two of the 2nd Battalion's Shermans. The Germans also dominated the battlefield through their unchallenged air superiority and effective artillery. None of Alger's supporting artillery executed effective counter-battery fire against the Germans, which permitted unmolested enemy artillery to rain down on Alger's forces.[62]

Lieutenant Kurt E. Wolff, a German tank company commander in the 21st Panzer Division, watched E and F Companies, 2nd Battalion, and a tank destroyer platoon maneuver south of the town of Sidi Bou

Zid. He waited for the order to attack with anticipation and dread as his force was outnumbered two to one. Wolff got the go-ahead to attack and advanced on the oblivious Americans. He had to move fast to get his company within effective firing range before the Americans identified the Germans moving in on them. Before Wolff closed the distance, one American tank identified his force, but the constant artillery barrage kept the Americans confused and allowed Wolff to occupy favorable ground. While he did so, other German units appeared on his right and left flanks to tip the odds in their favor. Wolff and his tanks opened up and immolated fifteen tanks in a matter of minutes. "They were burning. The flames were red and the first billows of smoke over the field of battle were dark gray and black. Anyone who has ever been in a tank battle knows how the heart beats when the attack starts rolling."[63] Wolff's company had not sustained even one death and could not believe how quickly they had won. Wolff recalled his battalion commander "laughing like a boy, went from company to company asking us and himself too: 'Did you ever see anything like it? Did you ever see anything like it?'"[64]

In the annals of military history, the attack of 2nd Battalion, 1st Armored Regiment, stands as a complete failure. Alger's leadership and tactical prowess prepared his unit as best he knew how. In less than a day, Alger moved, prepared, and executed an attack to save stranded comrades. He employed a savvy formation that provided adequate spacing and protection based on the estimate of the enemy situation. Lack of airpower, poor artillery support, inadequate intelligence, and superior German tactics and numbers turned the tide against the Americans on 15 February. These factors operated outside Alger's control and only his eagerness might have prevented a more cautious approach to the town of Sidi Bou Zid. Alger had received mentorship from experienced commanders and serves as a prime example of the military culture exhibited by officers of the 1st Armored Division. Combat experience is not everything in determining an officer's performance in combat, and the training, mentorship, and leadership experience prior to combat can prepare an officer to lead effectively.

Both charts in this section attest to the overwhelming superiority of the German defense against Alger's attack, and the results reinforce the data presented in chapter 2. The Germans outnumbered American forces at least three to one in tanks, not including antitank guns. The open terrain of Tunisia determined that large cannon fire would dominate this battle; infantry, artillery, and airpower played very limited roles. German tanks and gun crews, not pilots, artillerymen, or infantry-

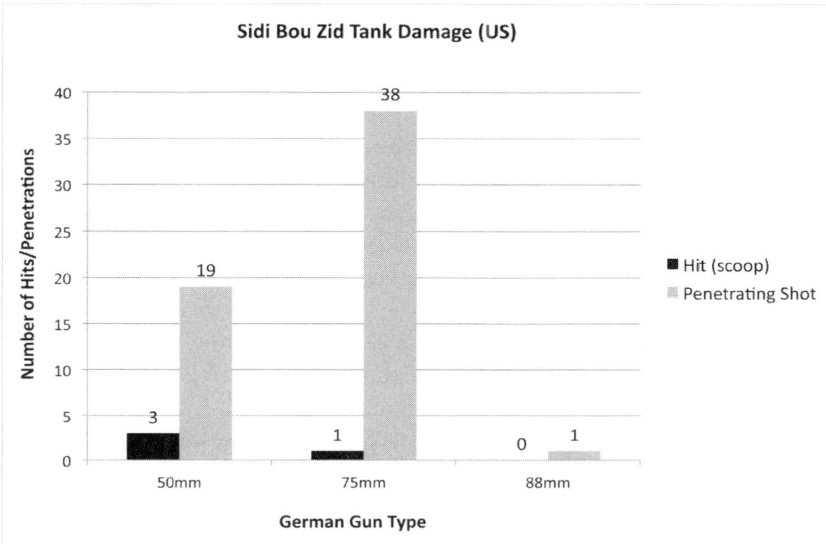

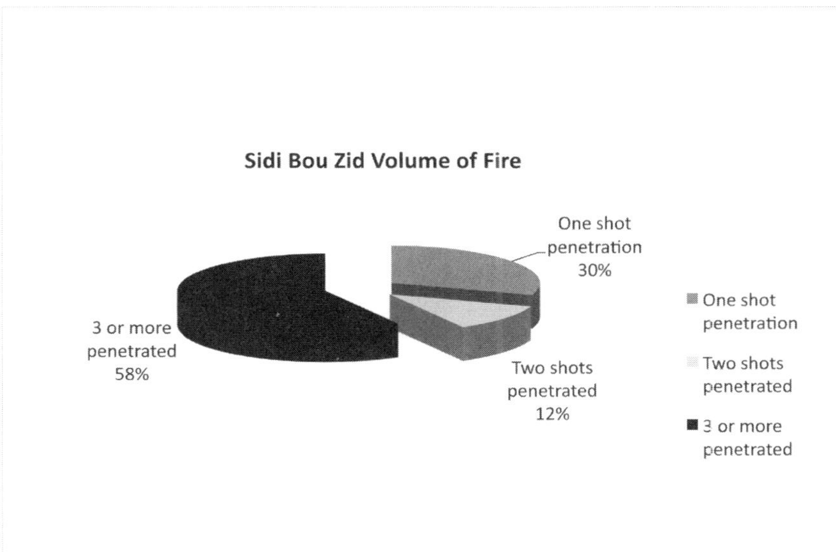

Sidi Bou Zid Tank Damage. Notice the high volume of penetrating shots as well as the number of penetrations compared to hits. From an untitled battle survey of the battlefield of Sidi Bou Zid conducted in early April 1943, Orlando Ward Papers, 1 page.

men, destroyed every single one of Alger's lost tanks. Even if Alger's battalion had had air superiority, a generous 10 percent attrition rate can be assumed against German armor, which would not have reduced the ratio enough to change the outcome. The multiple hits sustained by the crews of 2nd Battalion indicate that the tank crews fought degraded and bloodied until the tanks became uninhabitable or unusable. The vast majority of the German hits on the American tanks occurred on the side of the tank turrets and hulls, which indicates that the Germans enjoyed superior positions from which to engage the outnumbered Americans. Battle experience mattered little in the decision of this armored fight. No matter if Alger, Hightower, Waters, or Patton himself led this battalion, it seems doubtful that anything but a German victory was assured.[65]

The Germans had achieved a lopsided victory against a much smaller force against first Hightower and then Alger. Despite two days of heavy fighting, the Americans had no idea of the size of the force that awaited them in the vicinity of Faid Pass and Sidi Bou Zid. The Germans had braced for a serious American counterattack that manifested itself in Alger's assault. In two days of fighting against Hightower and Alger, the Germans had destroyed almost one hundred tanks and inflicted almost 1,600 casualties and POWs. Generals Fredendall and Ward deserve much of the blame for the poor commitment of these resources as well as the two stranded infantry battalions that ended up as residents of Poland and Germany for the remainder of the war. After Alger's fight, Ward had no idea what had happened and commented, "We might have walloped them or they might have walloped us."[66] Later in his diary, he wrote, "Alger was more or less on his own."[67] These two comments reveal an ineffective leader who failed to grasp what was going on around him.

Fredendall eventually authorized Ward to move Combat Command B under Robinett south to establish a blocking position in the vicinity of the Kasserine Pass. Unfortunately, Combat Command B, coupled with Alger's 2nd Battalion, might have made the difference at Sidi Bou Zid, and General Fredendall should have released it back to 1st Armored Division control long before the Faid Pass fight. Most commanders believed that the Germans would attack through the Fondouk Pass to the Ousseltia Valley to Maktar. The bulk of the Allies held the terrain in this vicinity, and Combat Command B did not need to remain so long in Maktar, due to the presence of the entire 1st Infantry and 34th Infantry Divisions.

Robinett's Combat Command B in addition to numerous U.S.,

British, and French units fought successive defensive engagements to stop the advancing Fifth Army and Rommel's units just west of Kasserine Pass. Both Allied efforts and the threat of the British Eighth Army in Libya eventually halted the Axis offensive in central Tunisia, and Rommel withdrew back to the east to consolidate his position, being squeezed from both sides. Lieutenant Colonel Henry Gardiner's 2nd Battalion, 13th Armored Regiment, participated in these hasty defensive engagements.

Stopping the Bleeding at Kasserine

A well-seasoned veteran after the landings and the race for Tunis, Gardiner organized his mixed battalion of Shermans and Grants in the Kasserine Pass to prepare for the Axis vehicles. Shrewd leadership existed throughout the unit: all three company commanders conducted a thorough reconnaissance of their fighting positions as well as withdrawal routes, the soldiers employed the local mud and clay to deftly camouflage their olive-drab green tanks, and Gardiner ensured his vehicles had hull-defilade positions and orchestrated a battalion volley of fire against the attacking Germans. This action differed from Alger's attack in that it was a defense against two panzer divisions instead of an offensive engagement. The numerical superiority of the Germans ensured them victory, but it would cost them more this time in terms of losses. The lone battalion defending against the Germans reflects the piecemeal employment of armored units so prevalent throughout this two-week period.

Gardiner initially counted thirty-five enemy tanks advancing toward his battalion. As the friction of combat grew around him, he lost count, but remembered that the number of German tanks kept increasing. The efforts of the 2nd Battalion initially stopped the German assault cold, but Gardiner recognized that no friendly units held his flanks, and the more numerous Germans began to outmaneuver his battalion. In addition, he received no artillery or air support that he constantly asked for during the battle. Gardiner had planned for this isolated tank battle, and his companies conducted an orderly withdrawal that incorporated overwatching fire by two companies while one displaced to the rear over preestablished routes. The wadi complex allowed for excellent cover, and some German tanks tried to disrupt the withdrawal. Gardiner's tank took two hits; the tank burst into flames, killing the driver; the rest of the crew escaped, and Gardiner had lost his second tank in four months. Gardiner personally extricated a wounded crew member and placed him on a functional tank for medical evacuation. He then made a 30-mile cross-country journey to rejoin his battalion after its successful withdrawal.[68] The valiant stand cost the 2nd Battalion ten tanks, but stymied the Axis offensive long enough for Combat Command B to establish another line of defense west of the pass.[69]

From the defeat at Longstop Hill until the frantic hasty defense at Kasserine Pass, the 1st Infantry Division suffered from the same piecemeal assignments that riddled the majority of General Fredendall's II Corps. Battalions from all the infantry regiments saw action defending eastern Tunisia with the British V Corps and French XIX Corps from the German Fifth Panzer Army's attempts to control the passes granting access to the Tunisian ports.[70] When not fighting these smaller actions, the battalion leadership spent time integrating new replacements flowing through the 1st Infantry Division training camp at Guelma, Algeria. The deteriorating situation at the Faid and Kasserine Passes required the commitment of RCT 16 and RCT 26. In mid-February, these units were thrown piecemeal into the fray and initially received their orders directly from the II Corps Headquarters. Cutoff battalions such as Gerald Kelleher's 1st Battalion, 26th Infantry, fought desperate actions in a fighting withdrawal to slow the German advance. In six hard days of fighting, Kelleher lost 8 officers and 235 soldiers, his operations officer, and his command post, and escaped after being captured.[71] As Rommel's Kampfgruppe DAK made its way through Kasserine Pass for the Bou Chebka Pass, the 1st Infantry Division coordinated a defense and counterattack that blunted the German attempts to control Tunisia. Both the 1st Battalion and 2nd Battalion, 16th Infantry, integrated field artillery and armor from the 13th Armored Regiment to establish strongpoints in and around the key terrain of the Bou Chebka Pass.[72] The battalion leadership of RCT 16 often cross-talked with each other about how to employ each others' units in the heat of battle. In one example, Lieutenant Colonel Joe Crawford's 2nd Battalion requested that the 3rd Battalion counterattack to seize Hill 812 overlooking the Bou Chebka Pass.[73] The coordination resulted in the seizure of the hill and recovery of field artillery guns lost earlier in the fight. Just like the other rifle battalions in the 34th Infantry, the 1st Infantry Division battalion leadership usually fought their units with little guidance or coordination.

The 1st Infantry Division's battalion leadership acquitted itself well throughout the frantic melee of the Kasserine Pass battles. The rapidly formed defenses controlled key terrain with well-coordinated artillery and armored support. These defenses convinced the Axis to call off their last offensive in North Africa. For the majority of the battalion leadership, this was their first encounter with synchronized mechanized formations on the offensive. Against some of the most seasoned and lethal German formations, the battalion leadership had displayed inspired and competent leadership even though much of their command structure and operational synchronization remained muddled.

The battalion leadership of the 1st Infantry, 34th Infantry, and 1st Armored Divisions endured significant hardships in February 1943. The defense of the Tunisian passes exposed the poor operational leadership of II Corps especially in regard to the ad hoc employment of U.S. forces. Catastrophic failures such as the capture of Van Vliet's 3rd Battalion stand as one casualty to the shortsighted leadership of Major General Fredendall.[74] Regardless of the errors committed outside of their control, the battalion leaders acquitted themselves with mature, courageous, and calm leadership no matter the odds. Waters and Gardiner had plenty of battle experience, while Hightower and Alger had experienced only air attacks and some minor reconnaissance skirmishes. Two of the commanders had tanks shot out from under them, and the other two ended up as prisoners of war. They employed their tank destroyers in the same manner, used sound fire and maneuver, and displayed courage throughout their engagements.[75] In every engagement, the German forces maintained significant advantages in numbers and aerial observation that precluded victory for the Allies based on ratios. Only better intelligence at the theater level and adept battle command from the division to the army level could have changed the outcome. Rifle battalion leaders ably inspired and led their infantrymen against enemy armored formations without close air support or significant numbers of tanks and tank destroyers. No discernible "greenness" among these leaders or their immediate subordinates surfaces despite overwhelming odds.

6

A Greenless Spring to End Tunisia

Bleeding from the face and with his field jacket punctured by two bullets, Major General Orlando Ward stumbled back into his command post early in the morning of 25 March. He had returned from an attack on the Maknassy Heights under direct orders from an impatient and frustrated Lieutenant General George Patton, the newly anointed U.S. II Corps commander. Patton shrieked the day before over a field phone: "I want you to get out there and get that hill. You lead the attack personally. Don't come back till you've got it."[1] Ward had led the entire 2,000-man 6th Infantry Regiment against the dark hill mass, Djebel Naemia.[2] While leading two companies of the 2nd Battalion, 6th Infantry Regiment, to the last of three ridges short of the objective, Ward had failed despite his personal courage. "Damn it, men! You're not going to let a 51-year-old man run your tongues out? Let's get up that mountain."[3] The infantrymen dug in 1,000 yards from the third ridge, but the 1st Armored Division did not completely control the heights above Maknassy until the Germans evacuated their positions on 7 April to consolidate their forces in northern Tunisia. The failure of Ward's division had severe repercussions.

Operational Leadership Challenges

Patton's dislike of Ward's personal style and his, undeservedly, negative opinion of the 1st Armored Division ended Ward's command. Major General Ernest Harmon, former commander of 2nd Armored Division, replaced him. The failed Maknassy operation sealed Ward's fate in Patton's eyes despite his own admission after he viewed the abandoned German positions on the heights that "I don't wonder that we took so long to take it. I wonder that we ever drove them out."[4] The

Germans had carefully prepared fortified antitank gun and machine-gun positions in the hard rock of the hills with explosives and pneumatic tools and were led by the talented Colonel Rudolf Lang, who fought on every major front and survived the war. His defense, consisting of approximately 350 soldiers, eight Tiger tanks, nineteen smaller tanks (Mark III and IV), copious amounts of artillery, and one dual-purpose 88mm gun, received the label of "German Thermopylae" in the German Fifth Army's logbook.[5] Ward's failure to take the heights on the first day he occupied the town of Maknassy, 21 March, stands out as his only misstep. In Ward's defense, First Army had ordered II Corps to maintain a defensive posture while the rest of First Army and Eighth Army attacked, and Ward received no guidance from Patton. Regardless, the relief of Orlando Ward supports earlier claims of flawed leadership at the corps level.[6]

While Fredendall exhibited almost no redeemable traits as a commander of mechanized forces, Patton was overly impatient and issued highly questionable orders to Ward. In one instance he suggested to Ward that getting more officers killed by placing them in exposed positions would bolster the morale of the enlisted men. He also ordered Ward to lead an infantry assault on a hill because he thought Ward did not possess personal courage. The politics of generals is not the focus of this book or this chapter, but the distant micromanagement of Fredendall and the manic aggressiveness of Patton illustrate the mismanagement of men and material on the battlefield. The complexity of division, corps, and army command requires talented leaders and experience. Men such as Eisenhower, Bradley, Patton, and Montgomery enjoyed long periods of time at high levels of command throughout the war, and they cut their teeth in the campaign for North Africa.[7]

Patton epitomizes the steep operational learning curve that U.S. Army generals started on with the North African campaign. While Patton was certainly a vast improvement over Fredendall, he still had to master the nexus of the tactical and operational levels of war. His uneven record as the II Corps commander is a prime example of the trajectory required to produce proficient commanders at the corps and army levels. Without this prior shaping experience, his Third Army's impressive northern pivot during the Battle of the Bulge seems unlikely. Moreover, II Corps had three commanders from February to the end of the Tunisian campaign in May. These short operational command tours identified one weak commander and provided critical experience for two more—Patton and Bradley. The limited command durations also reveal how heavily the rotating corps commanders relied on their tactical leaders, specifically at the battalion level. The corps commanders could not establish robust standing operating procedures or staffs attuned to their preferences. They also had precious little time to make an impact on the formations under their command.

The battalion leaders in the MTO did not require nor did they have the luxury of such an extensive education or time period of battlefield experience. The interwar army had prepared these battalion commanders enough to lead their units well. Their grasp of doctrine, history, and personal leadership qualities largely determined the outcome of fights, within reason. However, poor situational awareness and management of resources at the corps level and higher could set up disastrous conditions for the battalion commanders on the ground. Lieutenant Colonels Hightower, Moore, Van Vliet, and Alger became victims of this high-level ineptitude in the defeats at Faid Pass and Sidi Bou Zid on the 14th and 15th of February, respectively.

The Allied defense at Kasserine Pass stymied the Axis advance to Le Kef and ended Rommel's last offensive on 22 February 1943. The German offensive failed to obtain its stated objectives of a westward retreat of the Allied force and a significant reduction of its men and material in order to protect the north-south lines of communication and the Mareth Line. The Allied First Army spent about three weeks to set up the conditions for an advance through the reconstitution of ravaged units, transportation of key supplies (ammunition, fuel, and food), and surviving significant rains that made much of the Tunisian desert impassable for vehicles.[8] II Corps had a threefold mission that placed it in a largely subsidiary role due to British General Harold Alexander's diminished opinion of the Americans, especially in light of the disasters at Faid Pass and Sidi Bou Zid. Hence, II Corps's missions included attacks to force the commitment of Axis reserves, seizure of forward airfields in the hopes of tipping the balance of air superiority to the Allies, and seizure of Gafsa as a logistics supply point. The 1st Infantry Division took Gafsa easily, and the 1st Armored Division moved to the northeast headed to Maknassy. Through aggressive maneuver, Ward took Sened Station on 21 March with relative ease and quickly raced the next 20 miles to take the unoccupied town of Maknassy. The heights of the Eastern Dorsal that controlled the pass leading to Mezzouna loomed 5 miles to the east of Maknassy.[9]

On the morning of the 22nd, a reconnaissance patrol from Combat Command C (CCC) discovered both Djebel Dribica and Djebel Naemia, the two major mountains that made up the eastern Dorsals, "held in strength."[10] Other reports compiled later contradict this analysis, noting little activity, and Colonel Lang claimed that only about eighty German soldiers, part of Rommel's personal guard, and General Imperiali's Special Brigade 50 held the well-prepared position in the heights. Regardless of the force present at the pass, 1st Armored did not move to try to seize it until 2330 on 22 March. The series of attacks would fail, and Ward would go back to the United States. However, the efforts of II Corps achieved their objective since Kesselring committed Fifth Panzer Army's reserves to stop 1st Armored Division at the pass. He also committed the Army Group Reserve, 10th Panzer Division, to halt the 1st and 9th Infantry Divisions at El Guettar. The Axis held the positions at Maknassy and El Guettar to ensure the success of the Afrika Korps's retreat northward from the Mareth Line. The 1st Armored Division readied for the next phase.[11] The infantry divisions readied for their last big battle before the race to close the noose around the Tunis Bridgehead.

From El Guettar to Mateur: The Infantry Divisions Put It All Together

Both the 1st and 9th Infantry Divisions would remember the three-week El Guettar battle as the toughest of the entire war.[12] After Kasserine, the II Corps attacked eastward from El Guettar to retain pressure on the DAK as the British Eighth Army assaulted the Mareth Line. As the infantry divisions gobbled up more and more Tunisian territory, it forced the Axis forces to fight a tenacious delaying action anchored around the Tunisian ports. In the first phase, the 1st Infantry Division soundly defeated a vigorous attack by the 10th Panzer Division.[13] The 1st and 9th then counterattacked in an attempt to sever the enemy's access to the Tunisian ports. After this phase, the 34th Infantry joined the other two divisions under the II Corps to complete the final drive on Tunis. The battalion leadership within these divisions had to consistently employ aggressive maneuver to lead these formations against the Axis armor and intricate defensive positions carved out of the Tunisian hills. In order to overcome the stout German defenses, battalion leadership employed detailed reconnaissance to identify dominating high ground, coordinated their attacks to infiltrate as far as possible, and coordinated ground assaults with tanks and artillery. Simultaneously, division and RCT staffs got better at coordinating and supporting the maneuver of the subordinate battalions to maximize their effects on the battlefield.[14] When required, commanders exhibited extraordinary bravery. Major Charles Denholm led the 1st Battalion, 16th Infantry, in a night attack that culminated in a bayonet assault to seize Hill 523 that facilitated the 34th Infantry's seizure of the pivotal Hill 609 in the last week of April.[15] And Darby's always reliable rangers executed a daring night bayonet attack that netted 200 Italian prisoners in the vicinity of Djebel el Ank to begin the El Guettar campaign.[16] The battalion leadership suffered heavy casualties leading from the front.

No fight better illustrates the battalion leadership tenacity than El Guettar. The 9th Infantry Division had only two of its three assigned infantry regiments for the battle. Of the original six infantry battalion commanders, five became battle casualties. All three 47th Infantry Regiment battalion commanders, Major Herbert Turner (1st Battalion), Lieutenant Colonel Louis Gershenow (2nd Battalion), and Lieutenant Colonel John Evans (3rd Battalion), were either wounded in action or captured. In the 39th Infantry Regiment, Lieutenant Colonel Walter Oakes of 2nd Battalion was missing in action, and Lieutenant Colonel John Keeley of 3rd Battalion was evaluated as being in a state of "exhaustion," what we would now call post-traumatic stress disorder.[17] The loss of quality officers increased the "exhaustion-states" among soldiers, a clear sign of how dependent they were on their leadership.[18] Since 1st Battalion, 39th Infantry Regiment, was the division reserve and not committed until the 9th Division's attack had culminated, the 1st Battalion commander survived the fight.[19]

Since its scattered combat teams came ashore at Algiers and Morocco, the 9th and 34th Infantry Divisions had not trained, operated, or fought together until mid-March 1943. This was the same for all of the American combat divisions. In the wake of Kasserine Pass, Eisenhower had finally come to the realization that the scattershot approach did not maximize combat power, and he consolidated the complete divisions under II Corps. Regimental-sized units and below had gained combat experience since the TORCH landings, but division commanders and their staffs had done little more than coordinate movements of subordinate units. In the hills around El Guettar and outside of Mateur, the rifle battalion leadership of the three infantry divisions defeated a desperate enemy in the defense. They consistently displayed a penchant for incorporating lessons learned, leading from the front, and displaying a calmness under fire.

The 1st Armored Division moved to the Sidi Bou Zid area to rest and refit in mid-April. The Germans withdrew into a constricted bridgehead in the northeastern corner of Tunisia for their last stand against the Allies and languished in a logistical vacuum. The First

Army's plan of attack depended on the U.S. II Corps to seize the northern flank of the Germans, including the key cities of Mateur and Ferryville as well as the port city of Bizerte. Air superiority finally shifted to the Allies. On 20 April, the division left the Sidi Bou Zid area to finish off the Axis powers fortified in northeast Tunisia. By the 26th of April, it assumed responsibility for the southern flank, with 1st Infantry Division in the north. Their route demanded that the 1st Armored Division secure both sides of the Tine Valley and the exit at the northeastern corner called the "Mousetrap." The Tine Valley doglegs to the northeast, and numerous hills dominate the narrow exit from the valley toward Mateur. The 1st Armored Division would rely on an unlikely battalion commander to lead the way.[20]

Howze, Battle Plays, and the Breakthrough at Mateur

"I am no hero, but I am not a coward," wrote Colonel Hamilton Howze back in May 1943 in Tunisia in response to a question that he posed himself on whether he could endure combat duty or not.[21] Lieutenant Colonel Hamilton Howze had started his long army career thirteen years prior to his first combat command when he graduated from West Point in 1930. He had served primarily in horse cavalry units in the United States and the Philippines until assigned to the 1st Armored Division. After Tunisia, he would continue to distinguish himself throughout the Italian campaign as the 13th Armored Regiment commander. His post–World War II career included high levels of command such as corps (XVIII Airborne) and army (Eighth). More importantly, he commanded the U.S. Army Tactical Mobility Requirements Board (USATMRB), or "Howze Board" as it became known in 1962. This board produced the documents and policy that served as the foundation for the airmobile concept that gave rise to new units such as the air assault division.[22] As a cavalryman, Howze transitioned to mechanized warfare and led the vanguard of air mobility doctrine. But he needed to get out of Tunisia first.

Howze displayed some doubt prior to his experience in the vicinity of Mateur. Prior to assumption of command of 2nd Battalion, 13th Armored Regiment, he had served as the division operations officer, or G-3, for the 1st Armored Division. While a pivotal staff officer, he experienced occasional bombing, indirect fire, and assorted small-arms fire, and witnessed numerous enemy and friendly casualties. However, he did not consider his experience as a staff officer as combat experience and categorized most of his time as safe from the front line, engaged in the "continual grind" of staff work.[23] Bravery breaks down into two components: individual physical courage and responsible moral courage in leadership in combat. Regardless of his staff time, Howze possessed both components of bravery required for an effective combat commander.

On the 5th of April, Major General Ernest N. Harmon took command of the 1st Armored Division after Patton relieved Ward. Harmon brought his chief of staff, Colonel Maurice Rose, and his G-3, Lieutenant Colonel Lawrence Dewey, with him from 2nd Armored Division.[24] Howze ceased his employment as the G-3 and assumed the awkward job of executive officer of the 13th Armored Regiment. Unsatisfied with the position, Howze "talked his way" into a job that defies easy explanation. In short, he attached himself to the 81st Reconnaissance Battalion for about ten days and called in fire support for reconnaissance patrols and visited with advance reconnaissance patrols. While the division entered the area into Mateur, Howze remained with the 81st Reconnaissance Battalion at the vanguard of the approach.[25]

The 81st Reconnaissance Battalion, led by Lieutenant Colonel Charles J. Hoy, served as the organic reconnaissance unit of the 1st Armored Division. This unit primarily served as the eyes and ears of the division to determine enemy locations and find key terrain and road networks for the division's movement. The unit saw more of the North African desert and its hills than anybody else. Hoy's reconnaissance philosophy is expressed in the following statement: "Ordinarily, a reconnaissance unit will not fight for its information. This does not mean that it need not be aggressive. It takes 'guts' and drive to slip past the enemy, get behind him, and stay there transmitting information. But reconnaissance by fire should not be used promiscuously."[26] Hoy normally only had jeeps, trucks, and scout cars at his disposal, which forced him to maintain a passive posture. At Maknassy, he actually infiltrated scouts behind enemy lines and established observation posts. However, he knew that he must use force sometimes. When one German combat patrol compromised one of his observation posts in the Mouse Trap area, the lieutenant in charge radioed for the two tanks attached to his company, and they deployed over the knoll to kill, wound, or capture the entire patrol.[27] However, the constant need to fight for information escaped most commanders at all levels.

The most glaring void existed at the corps level, which did not possess any organic reconnaissance elements such as an armored cavalry regiment. If Patton had had this type of resource at his disposal, then he might have seized the terrain at the Maknassy heights prior to the Axis reinforcement. Even a seasoned leader such as Hoy, after intense reflection on his combat time, did not realize how critical corps level reconnaissance elements would be.[28] Howze understood the dearth of combat power in the 81st Battalion. His time with them, although unorthodox, probably influenced his actions in early May when he eventually took command of a tank battalion. He said, "reconnaissance patrols must be told about where enemy may be encountered."[29] Without constant information from aerial overflights or corps reconnaissance assets, the "about" had to come from survivable and lethal weapons platforms: tanks.[30] Lieutenant Colonel Howze's actions and coordination with the 81st Battalion reflected this attitude. Dismounted observation posts often took very long to establish, and a scout car did not last long against any German tank or antitank gun. From the record, it is clear that the 81st Battalion succeeded on the small-unit level, but was less effective at painting the big picture for commanders.[31] Howze realized this, but too late in the campaign to have any real effect beyond facilitating his own advance to the Mediterranean Sea.

Standing in contrast to the underdeveloped American intelligence doctrine, the Germans had consolidated the reconnaissance assets of the 15th and 21st Panzer Divisions in a composite *abteilung*,[32] an understrength brigade in American terms. The Panzer Army of Africa retained operational control over this robust organization, with good results.[33] The wide valleys, choke points, and desert terrain of North Africa demanded a fully resourced reconnaissance unit that could operate at the corps or army level. Unfortunately, the American division reconnaissance squadron remained heavily dependent on the support of the tanks of the divisions' armored regiments, especially the light tank battalions.

On the night of the 3rd of May, Howze settled in for a fitful rest while 75mm shells whistled in from the eastern side of the heights outside of Mateur every ten minutes. Due to his familiarity with the terrain and at Hoy's insistence, Howze conducted a reconnaissance in force on the 5th of May with a tank company and a platoon of assault guns, although to no avail. Howze returned to division headquarters after his action with the recommendation that an entire battalion of tanks could take the hill the next day. Colonel Clarence Benson took over Combat Command B (CCB) after an accurate artillery barrage had badly wounded General Robinett that day, and Howze returned to assist his new commander. Howze volunteered to coordinate the artillery support for the attack based on his knowledge of the terrain and the fact that the artillery observer for the tank battalion oozed inexperience. The attack, led by Lieutenant Colonel Gardiner of the 2nd Battalion, 13th Armored Regiment, culminated in a resounding defeat. The battalion attacked in three prongs with the sun rising in their faces and with no artillery support since Benson rejected Howze's offer of assistance with the artillery coordination. The battalion lost fourteen tanks, and Gardiner, the battalion commander, had to abandon his burning tank for the third time in North Africa and joined the ranks of the missing in action until late that night. To make matters worse, a tank platoon remained trapped on the west side of the hill. Howze told the platoon leader to stay put lest the tanks come under heavy enemy antitank fire. Benson denied his request for another attack. After an unexplained delay, Benson gave Howze command of the 2nd Battalion at 1300 hours in light of Gardiner's unaccounted status and instructed him to attack at 1600 hours. Howze informed Benson of the condition of the 2nd Battalion and warned him of the high cost of an attack this late in the day.[34]

Under these unpromising circumstances, Howze seized control of his first combat command and began preparations for a penetration that would continue for three days and end with total German surrender in Tunisia. Immediately, Howze called forward more ammunition for the empty racks in his tanks. Having seen the destruction wrought on the Shermans and Grants earlier by the well-camouflaged antitank guns across the valley, he then established a firing line on the ridge with a company of tank destroyers and a platoon each of mortars and assault guns to eliminate suspected positions.[35] The crew of his new tank gave Howze some much-needed instruction on radio use as well as other critical tank commander functions. Simply because they sat idle, Howze managed to secure the support of three battalions of the division's artillery support, numbering fifty-four guns! More importantly, the division artillery officer and Howze developed a sophisticated fire support plan that targeted suspected AT (antitank) positions and employed a heavy smokescreen of white phosphorous to isolate the right flank of the attack by obscuring the view of the heaviest concentration of German antitank guns.[36] Within three hours, Howze's task force sat poised to launch up the valley, and the fear of the attack washed over him. He gave a most vivid description on the subject of the fear a man feels prior to anticipated combat: "It is a monkey's paw that squeezes your liver in a heavy grip."[37]

Howze's force faced a potent assortment of antitank guns. The Pak 38 and 40 probably made up the bulk of the force, while assorted versions of the infamous FlaK[38] 88 or Pak 18, 36, or 37 were present, but in much smaller numbers than reported by American soldiers. Often, on the receiving end, every antitank gun looked like an 88mm, but they existed in far fewer numbers than the Pak 38 and 40.[39] Actually, the Pak 40 carried the lion's share of the antitank work for German forces. Both the 38 and 40 could be moved by a gun crew without a tractor or horses over small distances and solid ground, while this was not possible with the much heavier FlaK 88mm. German crews struggled to move larger guns unless they had tractors to pull the guns into position. The Pak 38, while boasting only a 50mm shell, could still penetrate all Allied armor, even the Soviet T34, until the tungsten carbide of its shell's penetrating core became too scarce for armor-piercing rounds. One would surmise that the rough terrain of the southeast of Mateur favored the use of the Pak 38 and 40. Regardless, all three types of guns had enough velocity to penetrate M3 Grant and M4 Sherman tanks, and only the 37 mm Pak gun proved inadequate except against half-tracks and light tanks.[40] The paucity of tanks encountered by Howze's force and the vast number of AT guns suggest that the tenuous logistics of the Germans precluded a healthy flow of fuel and maintenance parts for easy tank movement. Furthermore, German tank formations occupied positions that maximized their mobility to blunt Allied armored thrusts.

By May 1943, the German army had mastered the employment of antitank guns. In fact, they preferred to destroy Allied tanks with well-positioned and camouflaged AT guns, based on the higher survivability of an AT position over a tank.[41] In the compartmentalized terrain of eastern Tunisia and the Tunis Bridgehead, their penchant for this weapon system forced American attacks to be deliberate to achieve success. A rigorous training regimen compounded the AT gun crews' effectiveness. Each crew member usually completed an eight- to ten-week course that included holding gun drills in different terrain, firing at simulated tanks at ranges from 400 to 1,200 yards, and establishing AT positions. Crews had to meet the standard of sighting, loading, and firing their guns within thirty-five seconds and score an 80 percent kill rate when servicing targets.[42] Attacks against these positions required graduate-level synchronization of tanks, infantry, and artillery to isolate and destroy these lethal AT sites.

Howze led his trapezoid formation of tanks followed by a linear formation of tank destroyers over the ridge after ten minutes of artillery preparation. The effective smokescreen facilitated a rapid move to close the distance across the valley with the enemy guns. Howze, riding with

Table 6.1. German Antitank Gun Comparison

Antitank gun (mm)	Number in use 1 October 1944	Weight (pounds)	Effective range (yards)	Penetration power (mm)
75	4,805	3,136	1,000	102–130
88	578	8,000–9,660	1,500	130
50	no data	2,016	1,000	56
37	no data	970	400	49

Note: This table shows the prevalence of the 75mm over the 88mm on the battlefield by a factor of nine. Also, notice the significant weight difference between the 75mm and 88mm. The only true advantage offered by the 88mm was a greater range of 500 yards. Data from Werner Haupt, *German Anti-tank Guns*, 29–30, 39–40, and Werner Haupt and T. J. Gander, *German Anti-tank Guns, 1939–1945*, 63–64.

his head outside of the hatch as any effective tank commander must do, advanced through a hail of indirect fire that made hundreds of small pings and dings on his tank, and he found a hull down position to occupy for protection.[43] His tanks moved slowly once close to the AT positions and methodically overwatched each other's movements while they unleashed tank cannon fire on the Axis defenders. While he searched for an AT position, Howze looked down and realized that his tank sat in the middle of an infantry strongpoint strewn with fortified positions that contained infantrymen, machine guns, and AT guns. Howze scrambled for grenades, found only two, much to his frustration, and proceeded to toss one into a large foxhole and the other onto the belly of a German lying down and looking directly at Howze.[44] The hillside swarmed with scrambling German gunners and approximately eighty AT guns. While one company advanced, another company sat idle after a German 88mm disabled the commander's tank and Howze lost his radio transmitter. The company would not move despite Howze's frantic hand signals, and the attack sat on the edge of collapse.

In what he later called "the most difficult decision of my life," Howze dismounted his armored vehicle in the midst of intense artillery bursts and ran to each of the four remaining tanks of the static company.[45] He clambered up the sides of each of the tanks and shouted to each commander to follow his tank under the threat of court-martial. After an unhealthy dose of exposure to artillery in the open, Howze welcomed the comfort of his tank and moved forward with his new platoon to destroy an AT gun, which he claimed was an 88mm. The entire battalion advanced about 2 more miles and established a secure perimeter to organize itself and wait out the darkness until morning. While

Camouflaged German 88 Position in the Vicinity of Mateur. Photo 7-29, Robert E. Van Zant Collection, Photographs, General George Patton Museum, Fort Knox, Kentucky.

they feared an infantry attack and they had no infantry support to counter this threat, their fears were unfounded since most of the bypassed German infantry surrendered in large numbers to follow-on forces after the significant penetration.[46]

The trapezoidal formation utilized by Howze reveals a significant adaptation of armored doctrine that many officers of the 1st Armored Division employed throughout their tenure.[47] Commanders typically referred to this adaptation as "battle plays," and they used these pre-planned formations prior to, during, and after combat to exercise tactical control. A careful examination of the armored force field manuals published in 1942 makes no mention of these battle plays. *FM 17-10, Armored Force Field Manual*, does emphasize battlefield control as critical, but does not provide any real solution. A sample follows: "Simplicity of plans and formation, rigid adherence to the example of 'follow the leader' and reliable radio and visual communication are necessary for control at all times. During attack, control of small units is decentralized."[48] Vague doctrine required battle plays, which are part of an armored unit's standing operating procedures (SOP) in the U.S. Army

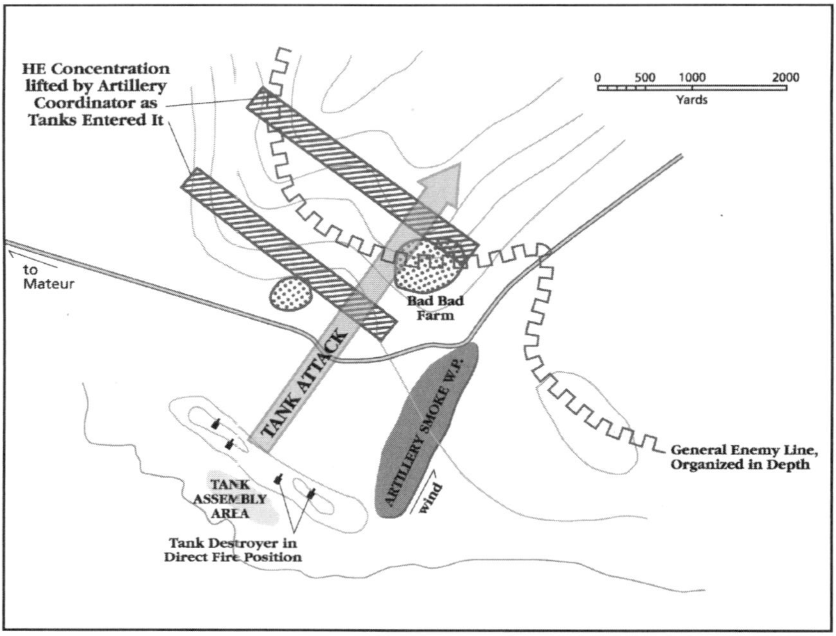

Lt. Col. Howze's Sketch of Mateur. Sketch drawn by Lt. Col. Howze describing the action. From Hamilton Howze, "Artillery Tank Support," Field Artillery Journal 33 (Oct. 1943), 779.

today, and General Ward paved the way with his innovative thinking.[49] Ward viewed these plays as a tool that commanders could use when they had to plan and execute an operation quickly and to adapt to the ever-changing conditions of the battlefield.

> It is not expected that these plays will necessarily be the solution on the battlefield, but it is believed that a thorough knowledge of them will furnish a sound basis for inspiration when the occasion arises. Our tactical thought is not to be rendered static by their adoption; they are 'speed plays,' for emergency use, or for other occasions where no time exists for a detailed study of the problem or for a rearrangement of our forces. A variation of these basic formations to meet the situation is a battlefield prerogative of every subordinate. A modified application of them will furnish the foundation for cooperative and effective execution.[50]

This battle play philosophy helped mechanized commanders visualize the employment of their units and, also, train their subordinates in the application of this new standing operating procedure. Throughout the North Africa campaign, the 1st Armored Division battalion and regimental commanders adapted battle plays to meet Axis forces.

Lieutenant Colonel Ben G. Crosby, commander of 3rd Battalion, 13th Armored Regiment, conducted several attacks throughout the 6th and 7th of May in support of Howze's main effort. Crosby had attacked on the morning of 6 May on the left flank of Gardiner, and his unit suffered a similar fate since it had attacked into the sun and could not identify the copious enemy AT positions. Later that day, he attacked again, just before Howze made his advance on the right flank. Under better light conditions, Crosby coordinated a three-prong assault on the western edge of valley. He deployed his tanks along a ridge and the base of the hill while he personally led a five-tank detachment to the top of the slope.[51] Crosby had employed "battle plays" when planning the advance and during the attack to adjust his vehicles. This allowed his force to destroy several AT positions as well as two Mark IV tanks. His use of battle plays indicates that he did not adhere rigidly to doctrine, but like all competent combat leaders, adapted it to the situation at hand.[52]

Interviews with Captain John McWatters, commander, Company G, 3rd Battalion, 3rd Armored Regiment, further show the institutionalization of battle plays within the 1st Armored Division. He acknowledged that Crosby's battalion routinely used these set plays throughout their time in North Africa. The quick change of missions, hastily planned attacks, and spotty tank-to-tank radio communication demanded the use of battle plays. During the division's withdrawal after the battle of Sidi Bou Zid, McWatters ordered three quick change of formation drills that neutralized German AT positions. The young commander's quick action prevented an antitank gun from inflicting casualties on an extremely vulnerable infantry half-track column from the 6th Infantry Regiment. The evidence suggests that battle plays had become part of the tactical vernacular and practice of the 1st Armored Division from the division commander and his staff down to the company level. This is just one example of tactical doctrine evolution, but the knowledge of it at all levels indicates that leaders coached, trained, and mentored their subordinates.[53] Even after the losses suffered at the battles of Kasserine Pass, Ward commented, "Our tactical teachings and technique have, I believe, proved to be sound."[54] This faith in the battle play and adapting to battlefield conditions paid dividends for leaders such as Howze, Crosby, and Benson as they tightened the noose on the Axis in Tunisia, even though their innovator had left command.

At first light, Howze moved his force to observe the valley through which they had just attacked and debated on how he could refuel and rearm his isolated force. Fortunately, the road lay open, and ammunition and gasoline flowed to the 2nd Battalion along with the 81st Reconnaissance Battalion and the remaining fifteen tanks of the 3rd Battalion, 13th Armored Regiment, which were headed north. The 2nd Battalion refueled and crept along the valley floor eastward. Howze's cautious movement paid off since his unit destroyed several more tank and AT positions as well as a huge cache of equipment in an olive grove.[55] The eastward progress through the valley consisted of careful movements and small actions with German AT gun positions and tanks. One of the tanks they encountered must have been a Mark VI (Tiger) tank since it took "many" shots to dispatch the vehicle at a range of 2,000 yards. One of the Tiger's dying shots removed the head of one of the company commanders, Lieutenant James Curry, who was respected as an extremely capable officer. Curry's death illustrates the potential cost of the dangerous but necessary practice of tank commanders exposing themselves to exercise proper command and control of their units.

On 8 May, outside of friendly artillery support, Howze continued his move to the north, this time to avoid the valley floor flush with tanks from the 15th Panzer Division. The armored column struggled up and down mountain slopes and across wadis that "strained the guts out of the tanks."[56] The unit labored for hours to climb each mountain and had to stop often to dismount tankers to place rocks in the deep gullies to make the route trafficable. The Germans had not fortified this side of the valley due to its ruggedness. Howze's men reached the end of the mountain range in full view of the German artillery and airfield at the end of the valley. The Germans opened fire, but it had no discernible effect on Howze's unit, and Howze placed the 2nd Battalion in a position to produce effective fire on the primary road. Howze halted any movement during the night and watched as the Germans torched their vehicles and supplies in anticipation of the attack that was sure to ensue at daylight by Howze's forty tanks perched in the rocky hills.[57]

On the morning of the 9th, Hamilton Howze sensed that the Germans had ceased to present an active defense. He jumped in a scout car and took off into the valley with his driver and a captured German machine pistol. Howze peeked through a hedge, and, before he knew it, 400 Germans brandished white flags and marched toward him with their hands up. His tanks along with 3rd Battalion caught up to their overwhelmed commander, and they headed down the road toward the coast. Tens of thousands of Germans surrendered, and Howze left his force at Bas-el-Djebel to handle all the Germans and their equipment. Howze continued to Metline with a platoon of five tanks and accepted the surrender of three German generals and 5,000 more men. The fight in North Africa ended for 1st Armored Division and Howze exactly one year from their embarkation from the United States to Ireland on 10 May 1942.[58]

Obviously, Howze was an exceptional soldier, but many officers had to make the same adjustments that he made with the mechanization of the army. His age, thirty-three, and experience mirrored that of his peers in similar command positions.[59] These commanders had to adapt their thinking and employ their grasp of this new doctrine on the battlefields in Tunisia at the user level. They adapted armor doctrine through improvisation and the use of techniques such as battle plays, and altered their comprehension of printed doctrine based on battlefield experience and training.

Commanders could and did overcome flaws in doctrine, unit organization, and the deficiencies of some of their equipment.[60] Many have argued that underdeveloped institutional doctrine as well as inferior equipment predetermined the American failure at Kasserine Pass. As pointed out in chapter 3, institutional doctrine does not exist only on paper, and a soldier's understanding of doctrine involves more than field manuals. The commander must possess the ability to think and to adapt his knowledge to solve a tactical problem. Furthermore, those who argue that inferior equipment was the fundamental U.S. problem usually compare only tanks, the 37mm guns, and tank destroyers. Almost every commander condemned the 37mm and tank destroyer doctrine in after action reviews and to ground combat observers before the Tunisian campaign ended, but these criticisms had little immediate effect on the ground.[61] I have addressed the issue of tank disparity earlier, but I will reiterate it here briefly. While the Sherman tank eventually replaced all the Grant-Lee tanks, its basic capabilities did not change except for the 76mm high-velocity cannon and some modest upgrades in the armor and powertrain. Mobility and lethality, not survivability, characterized the M4 Sherman tank even as more and more Panther and Tiger tanks prowled the battlefields of Europe. Arguably, the Sherman had its best chance of survival in North Africa given the paucity of German armor at the time. The quite opaque terms of combat experience haunted the American commanders throughout North Africa and suggests the following question: how is battlefield experience quantified, and when has a soldier shouldered enough combat to be considered battle-hardened or "war-wise"?

Cole and the Issue of Greenness

From 7 to 9 May, Lieutenant Colonel Lydon B. Cole, commander of 3rd Battalion, 1st Armored Regiment, led a task force to take the town of Ferryville and to clear the heights that bordered the southwest edge of Lake Bizerte, a large inlet that the Axis intended to defend well. His axis closely mirrored Howze's advance, only farther northwest. Cole's battalion had just come up from Mateur as the division reserve to exploit the momentum gained by Howze's exploitation. Although the Germans teetered on the verge of collapse, they still put forth stiff resistance at the key bridge crossing. His company commanders displayed some adept maneuver and initiative to eliminate or capture the ten tanks and antiaircraft guns that defended the bridge. Although not a decisive battle, it was combat nonetheless. Cole would later comment about these two days of fighting:

> Throughout the actions of the 7th–9th May, the tank companies were kept largely insulated from the hurry-up orders of higher authority, for which the battalion commander assumes full responsibility. On only one occasion in the three days (on May 7) was it deemed advisable to spur a company along. Otherwise, the tank companies, once they were committed, were permitted to set their own pace; they operated as they had been taught to operate since they became "war-wise," slowly, deliberately, thoroughly, and with a maximum of observation, time for effective fire and the maintenance of perfect and complete control throughout the chain of command at all times.[62]

The use of the phrase "war-wise" stands out, but it lacks a clear definition, although one can infer its meaning from Cole's usage. No combat event repeats itself exactly, but the disconcerting and undesirable effect of someone shooting at you has its own way of making a lasting impression on any soldier. Cole and his commanders had already experienced enemy fire by the time they rounded Lake Bizerte. Of course, it does not take a catastrophic immolation such as what happened to

Alger's battalion at Sidi Bou Zid to make soldiers war-wise. Units can still fight effectively for the first time even if not war-wise, but, as Cole points out, this baptism of fire produces more methodical and lethal soldiers. Even though he executed his first engagement with only four hours of planning, Howze's impressive smokescreen to permit his assault on AT positions highlights a soldier who had achieved his "war-wiseness" prior to command.

The other interesting component of Cole's comments is a reflection of the tremendous relief that he and his commanders enjoyed from the interference of their superiors' commands during the latter stages of the campaign. The irrational pressure inflicted by Patton on Ward at Maknassy stands out as prime example. Commands issued over the radio or via field phone that encourage people to move faster and be more aggressive usually do not yield favorable results, particularly in the vast array of choke points that dominated the terrain in Tunisia. In fact, commanders such as Robinett often advised their subordinates to avoid "rat racing," as it was known in the 1st Armored Division.[63] Howze, probably due in part to his senior position and assumed competence, enjoyed a free hand in his operations in early May.

A return to Lieutenant Colonel Hamilton Howze's actions on the 6th of May provides a telling glimpse into combat experience. Howze had never commanded a unit of this size before, nor had he commanded under combat conditions. While he knew some of the officers, he did not have time to train them or know them well with only four hours before an attack. He had never before fired and maneuvered to take an enemy position as all of his subordinates had done prior to him assuming command. His celerity in planning the operation indicates that he understood armored doctrine and combined operations well. He employed the well-known technique of battle plays to explain his plan of attack and control the movement of his subordinate units. His participation with the 81st Reconnaissance Battalion's collection efforts provided him with not only an appreciation for accurate reconnaissance, but also the actual information he needed for his upcoming attack. Since 1st Armored Division was not the main effort, air support in terms of close air support or aerial reconnaissance did factor into this decision making. He was fortunate that Germans did not possess local air superiority as they often did after the Americans landed in North Africa.[64] The most outstanding example of Howze's performance is the devastating effect that the use of artillery had on the Axis positions. Howze would later comment about the effectiveness of the artillery barrage and smoke:

> From my tank I could see the hill coming apart in front of me, while an excellent smoke screen guarded our right flank. The attack swept on without a hitch—though the enemy put down a fairly heavy concentration of large caliber HE, his AT guns short of the objective never stood a show. A few were overrun by tanks but the majority were not fired, for the gunners could not man them in the hail of HE. Some of our artillery landed close to, and occasionally even behind, our tanks, but it seemed sort of friendly. As the tanks crossed the ridge we encountered more AT guns and our artillery, which could not see us, lifted. But these enemy guns were hastily emplaced, and the tanks shot them up in short order.[65]

Even though Howze had wanted to wait until the next morning to attack, he managed to use his abbreviated timeline to achieve remarkable results given the circumstances. His combination of overwatching fire from tank destroyers, obscuration, and accurate artillery fire suggests an impressive grasp of the enemy situation.

But Howze definitely showed signs of fear, and he did not like it. He viewed this action as his first real test of combat and worried about how he would measure up to the challenge. Two distinct incidents stand out as turning points in his experience: in the first, he waited for his attack to begin with wrenching pains in his guts; in the second, he gathered up enough courage to exit a tank in the midst of a heavy enemy artillery barrage. He no doubt experienced other anxieties as AT, tank, and artillery fire swirled around him. Howze displayed courage in his use of grenades against enemy infantry that occupied positions literally feet away from his tank, who could have dispatched an exposed tank commander with relative ease. The habit of tank commanders riding exposed in their hatches requires vulnerable exposure to enemy fire, but it remains a critical task of leading armor formations.

Lieutenant Colonel Howze achieved the stamp of combat experienced in his first fight. The rite of passage of combat is critical, but how much is enough eludes clear definition. Moreover, it clearly varies from person to person. Exceptional leaders like Howze probably acquire an adequate sense of being war-wise rather quickly. However this may be, the issue of whether or not American commanders at the battalion level had shed their greenness was dead long before the final assault against the Axis in northeast Tunisia.

When accepting a German general's surrender, Howze met his adjutant, a captain who spoke perfect English from his upbringing in an international compound in Tokyo. His English served the necessary purpose of communicating to Howze exactly where all the units, equipment, and supply depots were located in order to facilitate an orderly surrender. Their conversation then turned to the captain's background in the German army. He told Howze that he had spent a long period on the Russian front, "where he said the fighting was much more intense and brutal than in Africa."[66] This statement reflects the harsh conditions and barbarity of the combatants on the Russian steppe, forests, and swamps on the so-called eastern front, but belies the desperation of the Axis in North Africa. The scale of operations in North Africa does not quite compare to the vast events in the Soviet Union at the time, but the breakdown of the Axis in Tunisia showcases an important element of combat experience as well as a loss of significant combat power. Approximately 250,000 Axis soldiers of Fifth Panzer Army, Afrika Korps, and Italian First Army making up Army Group Africa surrendered to the Allies in Tunisia, compared to almost 300,000 casualties sustained by the Second German Army and Italian Eighth and Hungarian Second Armies at the same time on the eastern front.[67] The German forces that held out to the end maintained discipline in their units, an offensive mind-set, and a high state of morale. Often poorly informed, they believed that they could hold out in Tunisia indefinitely.[68] The German armored formations had lost considerable numbers of experienced veterans fighting in North Africa, which, coupled with the woefully inadequate logistical system, contributed to the downfall of their foothold in Africa.[69] Keeping combat veterans alive and uncaptured to utilize their experience is critical to the success of an army. While Germany had already slipped away from its high-water mark of combat experience by May 1943, the American battalion leadership in North Africa had obtained that experience, and it would continue the fight in Sicily, Italy, France, and Germany.

7

The Last First Battle, Sicily

While the Allies closed the ring against the Axis in Tunisia, the 67th Armored Regiment executed a rigorous field exercise in Morocco to prepare for future operations in Sicily. Based on the United States II Corps's tangling with Axis forces since the TORCH landings, commanders at all levels seized the opportunity to train their formations. The 67th's exercise that began on 24 March included an ambitious agenda. Essentially, the exercise was a movement to contact—one of the most difficult missions, based on the vague conditions and enemy situation. The exercise's operations order placed maximum emphasis on realism to include a forty-hour continuous operation length with no plan for sleep, a simulated gas attack, live mines, and evacuation of simulated casualties. The 67th Armored Regiment received orders to move north to destroy the "Red (German)" tank and infantry task force (played by the 82nd Reconnaissance Battalion) moving south. Both forces moved and fought each other at night to represent the most challenging set of circumstances in a training environment. Rigorous field conditions such as this one indicate a climate of mature leadership focused on preparation for future combat.[1] The 2nd Armored Division's battalion leadership used its time wisely to integrate new soldiers in intense training based on the lessons learned by its sister armored division slugging it out in Tunisia.

Unlike other U.S. divisions on Sicily, the 2nd Armored Division almost never employed its battalions as whole units. To support the infantry battalions' assaults on beaches and the fortified hilltops in the mountainous terrain of Sicily, the 2nd Armored Division was a force provider and the Seventh Army reserve. The 2nd Armored served as the headquarters for the KOOL force—the floating reserve for the HUSKY landings. The KOOL force was really the 2nd Armored Division headquarters with one combat command (CCB) and the 18th Infantry Regiment standing by to sup-

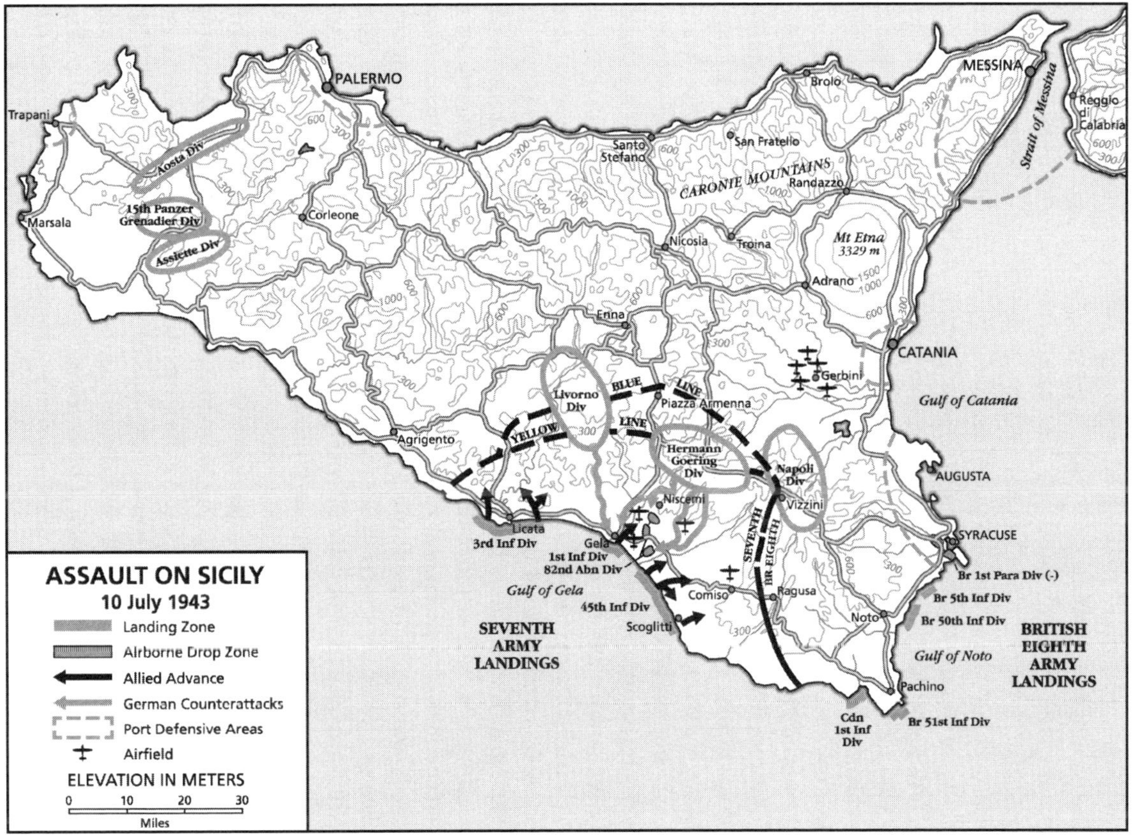

Overview of Operation HUSKY Landings. From *Sicily*, The U.S. Army Campaigns of World War II, Center of Military History Publication 72-16, no date, 11.

port the beach assaults. The remaining tank battalions provided all of their subordinate companies—one to each regiment—to support the RCTs of the infantry divisions. The lone tank battalion of the CCB, 3rd Battalion, 67th Armored Regiment, acquitted itself well as the 2nd Armored followed and supported Seventh Army's race to Palermo and Messina. Fuel shortages and defective track blocks severely hampered the employment of 2nd Armored.[2] However, the intense North African training paid off tremendously for its fragmented platoons and companies supporting the infantry battalions. The 3rd Battalion commander's actions at Gela confirmed that the tank outfit that "can shoot first and straightest" will win the engagement.[3] The 2nd Armored's tankers had practiced their tank gunnery every month since March. As in the last month in Tunisia, the Axis's adroit placement of the antitank guns presented a tactical challenge in the Sicilian mountains. The 2nd Armored Division's armored formations provided survivable platforms to suppress and destroy these lethal AT positions.[4] In its most memorable action, the 2nd Armored Division moved 60 miles in ten hours from the Belice River to Palermo. Many remembered that the action closely resembled the division's efforts during the Tennessee and Louisiana maneuvers of 1941 and thought that the maneuvers seemed more demanding than actual combat.[5] The 2nd Armored Division's battalion leaders oversaw the seven-month training regimen in North Africa that paid dividends for their tank platoons and companies in Sicily.

The Sicilian Campaign occurred in two phases: a massive amphibious operation and, then, a mountain campaign in pursuit of Axis forces to Messina. Throughout both phases, a combination of U.S. units with varying levels of combat experience battled for just over thirty days. After much negotiation at the Casablanca Conference in January 1943, the Combined Chiefs of Staff had settled on Sicily as the next objective in the natural progression of the Mediterranean strategy. The Allies recognized the island's significance for controlling the sea lanes and providing airfields for the eventual invasion of the Italian boot. The invasion plan called for the U.S. Seventh Army to land on the south coast of the island at the main towns of Gela and Scoglitti with General Bradley's II Corps, and Major General Truscott's 3rd Infantry Division, reinforced by CCA from 2nd Armored, assaulted Licata to protect the entire flank of the invasion. In addition to establishing an Allied bridgehead, the Seventh Army was to seize the airfields near Gela and seize the port of Licata. The British Eighth Army landed four divisions and one infantry brigade on the southeastern coast to capture the port of Syracuse. Both armies employed vertical envelopments to seize key terrain and intersections behind the beachheads. After the landings, the plan remained murky as to how the Allies would occupy the remainder of the island. The HUSKY landings achieved their objectives and defeated Axis counterattacks, and then the Allies slogged through a monthlong campaign battling an enemy fighting for its life.[6] Operation HUSKY was the last campaign conducted solely by division cadres that had consolidated before 1942, and therefore included a very high percentage of peacetime regular officers. It stands as the final test to analyze the combat effectiveness of battalion leadership that ably led their units across the rocky island. No matter what their level of combat experience was, commanders and ground observers all concluded that units which executed rigorous and thorough training in preparation for combat could defeat a veteran enemy in rugged terrain.[7] "There may arise a mistaken notion that much of precombat training is not effective because in the final analysis troops can learn only in battle. Such a notion has been thoroughly exploded and disproved by actual experience in Sicily."[8]

Thunderbirds Rising: Building Learning Organizations Prior to Combat

As the only division with no combat experience, and coming directly from the United States to the shores of Sicily, the 45th Infantry Division started at a disadvantage. Coming ashore between the seam of the U.S. Seventh Army and the British Eighth Army, the Thunderbirds drew a critical assignment, given the ferocity of the German counterattacks. The Thunderbird Division synchronized their amphibious landings well, considering the rough surf and wind conditions. The only significant setback came when elements of the Hermann Goering Division overran one battalion from the 180th. Within three days, all the RCTs captured the assigned airfield objectives even after tangling with the attacking elements of the Hermann Goering Division.[9] One battalion executive officer from the 180th Infantry Regiment, 45th Infantry, validated the importance of practicing techniques in training to maximize the effectiveness in combat with this comment about tank-infantry coordination:

> In the U.S., we always trained with what we had, and never had any training in cooperation with tanks. When we got into action over here in Sicily and we're given tanks to assist and support us,

> we actually didn't know how to use them or work with them. There must be real training for the infantry with actual tanks in realistic combat exercises. We could have done much better in several of the places over here if we had been given training in this before going into action. Only twice did we get real benefit or advantage in the use of attached tanks, and this wasn't the armored people's fault. We were just too unfamiliar with the proper way to use them.[10]

The 45th Infantry moved with the Seventh Army to quickly take Palermo in a rapid maneuver to the northwest corner of Sicily. From Palermo, the Seventh Army had only two roads to use in order to reach the town of Messina: the coast road, highway 113, and highway 120, which snaked its way through the heart of Sicily's Carnoie Mountains through Troina to Messina.[11] The Axis built this northeastern corner into a redoubt called the Etna Line (named after the volcano that anchored the line). The 45th Infantry fought eastward down 113 until the 30th of July, when it was relieved by the 3rd Infantry in order to prepare for the invasion of Italy.[12] For twenty days, the Thunderbirds fought along this narrow highway to dislodge the Axis from ridgeline after ridgeline. To infiltrate or assault mountainous terrain, the infantry battalions had to attack at night. Battalion commanders harped on their subordinate company commanders to obey simple principles when attacking. These principles consisted of controlling dominant high ground, coordinating assets, and attacking at night. Rifle company commanders often got consumed with maneuvering their organic platoons and forgot that they should use the weapons company, the regimental cannon company, attached armor, mortars, or artillery. One battalion commander commented, "In some cases they [company commanders] seemed to forget that they had heavy weapons in support."[13] When attacking Axis strongpoints in Sicily, battalion leadership proved its worth in ensuring that their companies employed the resources available to defeat the enemy and keep casualties as low as possible. American artillery was a prominent advantage that could have devastating results as the field artillery got better and the U.S. Army could mass them more easily along the shorter front in Sicily. In some cases, divisions dedicated one artillery battalion to each company against particularly tough Axis positions.[14] The battalion leadership's maturity and past training experience made these lessons clear to them, but they had to invest considerable time in their junior officers.

Rock of the Marne: Overcoming Friction at Licata

The Marne Division's landings along the Licata beaches—code-named JOSS—encountered higher levels of resistance than the Big Red One, 45th Infantry, and rangers did at the Gela beaches. Like the 1st Infantry Division, the 3rd Infantry Division built task forces around infantry battalions to overcome the Axis defenses on the southeastern Sicilian coast. Three aspects of the division's training program for HUSKY showcase the criticality of seasoned battalion leadership. First, the 3rd Division's staff became so consumed with establishing a joint headquarters to coordinate the landings that subordinate units mostly prepared their units for combat with little to no guidance or resources from the division headquarters. Second, the division did not even consolidate its forces for the mid-July landings until June 20th. Finally, the division provided 3,000 infantry and artillery soldiers as well as 160 combat-arms, company-grade officers as replacements for the losses suffered by the 1st Infantry and 1st Armored Divisions after the Kasserine Pass battles in late February.[15] Over the next five months, the division received seven sets of soldiers to replace these losses. Unfortunately, 85 percent of these soldiers had received only thirteen weeks of basic training.[16] A more telling statistic was their rifle qualification. Of the 1,416 soldiers who possessed a rifle qualification record, only 908 had actually qualified their rifles. Battalion commanders remained a constant throughout this turmoil to design the training to integrate these new soldiers into their formations prior to the landings. Eight of the nine rifle battalion commanders remained in position from February 1943 until the completion of the occupation of Sicily.[17] The 3rd Ranger Battalion, which served as one of the assault force battalions for the 3rd Infantry Division, had existed only since early May 1943. Plucking a cadre to form the nucleus of the new ranger unit, the battalion leadership spent eight weeks training the soldiers for the Sicilian landings.[18] The ambitious agenda included mountain assaults and amphibious landings.[19] In both the 3rd Infantry rifle battalions and the 3rd Ranger Battalion, battalion leadership built cohesive and effective combat teams on an abbreviated timeline and with new soldiers.

The field order for Regimental Combat Team 7's landings at Red Beach on Sicily was merely six pages with ten annexes, but the regiment published only two of these annexes. Nine days prior to RCT 7's departure from Bizerte, Tunisia, the battalion leadership of the three rifle battalions had this field order and two annexes—an operations overlay and a landing diagram and landing schedule—in their possession to prepare their units. They did not have the communications, air support, naval gunfire, or alternate plans to refine their planning. Major Everett Duvall, the 2nd Battalion Commander for the 7th Infantry Regiment, had five tasks assigned to his unit. These tasks were:

1. Advance in its zone of action with utmost speed, destroy all enemy encountered, seize objective L, and hold the enemy North and West of line OP.
2. Protect the right of RCT-7.
3. Establish and cover road block at W.
4. Destroy railroad vicinity of W.
5. Be prepared, upon accomplishment of mission, to attack and destroy by-passed enemy positions.[20]

In addition to these specified missions, Duvall was expected to integrate a tank company from the 66th Armored Regiment if directed by the RCT 7 Headquarters. The 2nd Battalion also had to clear all structures encountered during the assault and not take the field order or any annexes ashore. The minimalist nature of the RCT 7 Field Order for Operation HUSKY remains consistent with other orders such as ones issued by the 1st Infantry Division during the attack against Oran, Algeria. Besides the similarities, the ubiquitous use of this format further highlights the capabilities of battalion leadership to prepare for and execute operations with negligible direction from above.[21] Duvall and the other battalion commanders translated operational guidance into tactical reality.

The 3rd Infantry Division's assault plan involved a complicated double envelopment, placing RCT 7 in the most difficult western position. Besides the difficult surf conditions, Red Beach was also the most heavily fortified of all the Sicilian beaches. A thin beach of only 8 to 20 feet in depth with 60-foot cliffs and no beach exits except three paths on the west end, it presented the most challenging objective for the Marne Division. Lieutenant Colonel Roy Moore's 1st Battalion was designated the assault battalion to land first on Red Beach. Surf conditions and poor landing craft handling delayed the 1st Battalion's assault. This delay caused significant confusion when Duvall's 2nd Battalion began its approach for the landings. As the LCIs (landing craft infantry) bearing the 2nd Battalion approached the beach, they turned about when they did not see 1st Battalion's LCVPs (landing craft vehicle personnel). The flotilla commander finally determined that the landings were back on schedule when he noticed several LCVPs moving toward Red

Beach. He then decided to hold the LCIs for twenty-five minutes so Moore's First Battalion could land and clear the beach. Once it started to become light at 0415, the 2nd Battalion's LCIs headed for the center of Red Beach and landed before 1st Battalion. Now being the assault force and under constant enemy machine-gun and artillery fire, Duvall organized his men to get off of the beach to seize their assigned objectives as well as defeat an Italian counterattack against the railroad station at San Olivia. The third battalion under Lieutenant Colonel John Heintges suffered the same fate as the 2nd Battalion and had to overcome the beach defenses to get to its objectives.[22]

When RCT 7 came ashore on 10 July 1943, the three battalions were supposed to land in numerical succession and seize their assigned objectives. The poor synchronization at the regimental headquarters and higher, weather, mishaps, and planning shortfalls created significant turbulence for the battalion's landings. In anticipation of daylight landings, RCT 7 planned to land its 4.2-inch mortar squads before the assault waves to provide smoke obscuration. Poor coordination by the RCT 7 HQ delayed the mortar landings and any hope of smoke. Successive landings felt the impact of this mismanagement. RCT 7 could not silence an enemy 47mm gun position that eventually destroyed two LCIs. It took approximately two hours to level naval gunfire and artillery to silence the gun and protect the rest of the landings.[23] The two destroyers meant to provide naval gunfire support on Red Beach had collided three hours earlier.[24] The integration of both naval gunfire and close air support could have provided more timely fires, especially against strongpoints that the infantry was not equipped to destroy. None of the landing forces had the ability to direct air support, and no aircraft were designated for them in any case. The limited range and capabilities of the wireless radios induced significant confusion into RCT 7's landings. Observers even commented on the stretched communications capabilities with the damning statement that "wireless was a totally unreliable means of communication; and that we should do much better to regard it as a luxury and resort to other means, including visual, for vital signals."[25] A whole host of factors significantly altered the plan that Duvall had developed for his battalion on Red Beach. Nevertheless, his unit and Lieutenant Colonel John Heintges's Third Battalion quickly adapted to the circumstances to still meet their assigned objectives.

RCT 15 grappled with the same issues on Yellow Beach, but the 3rd Rangers on Green Beach and RCT 30 on Blue Beach managed better success with preplanned naval gunfire and pre-landing of mortar platoons for smoke. At all four beaches, the dogfaces of the 3rd Infantry Division faced machine-gun and rifle fire that individual battalions defeated.[26] Despite the light opposition, all battalions had carried out an assault on a moonless night with no preliminary softening of the beaches by naval or air power.[27] Of the eleven "rules" laid down by the 3rd Infantry Division commander for the landing of the regimental combat teams, number four dictated that each regimental commander would "land personally" with the second or third battalions and locate his headquarters in an LCI that had command and control capabilities.[28] This explicit rule reveals two things. One, division commanders expected their leadership to place themselves in the right spot on the battlefield at the right time. Two, senior commanders recognized that when leaders made decisions at the critical location in a battle, they could reduce or overcome the friction of warfare. Other lessons-learned documents and observer reports confirmed the requirement for commanders to be with their troops. Battalion commanders not only had to set the example, but to provide adequate command and control in light of spotty communications with higher headquarters.[29] While Truscott felt the need to spell out this order for his regimental commanders, it is likely he assumed that his battalion commanders would land and lead their units. The three battalion commanders for RCT 7 embodied this expectation through their will to achieve their missions no matter the conditions on Red Beach. After the landings, the 3rd Infantry Division raced to Palermo and eventually replaced the 45th Infantry along highway 113 to secure the northern coast. In its advance to Palermo, the division's intense training showcased its dividends when it marched 100 miles in just over three days' time.[30] During the attack along the coast, the division executed two more amphibious landings to bypass stout Axis defenses. In the end, the division became the first elements to reach Messina.[31]

The Fighting First at Gela

Of all the divisions in the U.S. Army, the 1st Infantry had accumulated the most combat experience by the time its units disembarked from landing craft onto the island of Sicily. The battalion leadership had previously proven their worth from Oran to Kasserine to El Guettar. One of the battalion commanders, Bowen, had already risen up from battalion command to take command of the 26th Infantry Regiment in April 1943. Many commanders had already earned valorous combat awards or escaped Axis capture, such as Kelleher at Kasserine and Denholm at Mateur. In fact, Patton selected the tired division for the HUSKY invasion over an untried 36th Infantry Division that planners had originally tapped for the task.[32] The 1st Division's success stemmed mostly from the brave and competent battalion leaders who could execute their missions with little guidance or operational support. A discussion of the Fighting First in Sicily highlights the battalion leadership's ability to integrate new soldiers into their formation to build combat-effective units. Even though the 1st Division's landings did not encounter significant resistance, the subsequent counterattacks by the Hermann Goering Division and Livorno Division severely tested the units in the first few days of securing the bridgehead.

Unlike the 2nd Armored Division's longer training program, the 1st Infantry Division had just less than two months to integrate over 2,000 replacements and practice its amphibious landing techniques near Arzew, Algeria. The training with the new landing craft and soldiers paid off. The 1st Division came ashore at Gela, secured the Ponte Olivo airfield, and reduced Italian strongpoints. However, the Axis forces had positioned their armored units close enough to the beaches to execute counterattacks against the beachhead. The Italian Livorno Division and the German Hermann Goering Panzer Division launched counterattacks from 10 to 12 July, mostly against the 1st Infantry Division and elements of the 82nd Airborne Division and the 45th Infantry Division. Three of the 1st Division's battalions absorbed the brunt of the Axis counterattacks. In one instance, General Paul Conrath's Hermann Goering Division hit Joe Crawford's 2nd Battalion, 16th Infantry, as they defended positions overlooking the Niscemi Road. Although the German tanks bypassed Crawford's position, he effectively controlled his unit to remain in their foxholes to defeat the supporting German infantry. As Conrath's tanks continued down the road, they ran into Lieutenant Colonel Denholm's 1st Battalion, 16th Infantry. Denholm used his forward observers' ability to call in naval gunfire and antitank fire to stymie the offensive. Denholm was wounded during the ferocious attack when he attempted to man one of the antitank guns before the naval gunfire landed.[33] In a similar incident to Crawford's, Lieutenant Colonel John Corley's 3rd Battalion, 26th Infantry Regiment, engaged elements of both enemy divisions near the Ponte Olivo airfield at night. The German tanks bypassed the dug-in infantrymen, but Corley's well-coordinated defense denied the tanks their infantry support. As the tanks reached the Gela plain, direct fire from the 1st Infantry's artillery units and naval gunfire tore into the unsupported armor.[34]

The successful protection of the beachhead stemmed from the battalion leadership's ability to keep its formations in position despite the terrorizing attacks of the armor. The unit's destruction of Axis infantry prevented them from supporting the final armored assault against the Allied beaches. The disciplined actions stand as a testament to the importance the battalion leadership placed on training and leadership throughout the training program for the HUSKY landings. Immediately following this successful defense, the 1st Division went on the offensive, mostly at night, for the next five days to exploit their success. In similar circumstances at Gela, RCT 26 endured some tough fighting against desperate German armor attempting to retreat east. This offensive spoiled another Axis counterattack. For the next three weeks, the 1st Infantry stormed successive Axis positions through the Etna Line along highway 120 until reaching Troina—the Axis keystone to the Etna Line. After six days of the hardest fighting the division witnessed in Sicily, the position fell.

Similar to the 1st Infantry's training, Lieutenant Colonel Darby trained his Force X—consisting of the 1st and 4th Ranger Battalions, the 39th Engineer Battalion, the 1st Battalion, the 531st Shore Regiment, and the 83rd Chemical Battalion—in less than a month for the invasion. Attached to the 1st Infantry, his force was to destroy coastal defenses northwest of Gela to protect the landing beaches, which was similar to his first combat mission in Arzew, Algeria. Darby ably led this regimental-sized task force to achieve its HUSKY objectives. Darby later credited the unit's success to the month of training he dedicated to the integration of his new soldiers, especially the newly created 4th Battalion.[35] Incorporating lessons learned, Darby had attached significant firepower improvements to his organization such as the 4.2-inch mortars and 37mm antitank guns. Earning a Distinguished Service Cross for his personal destruction of an Italian tank, he continued to prove his worth as an effective and brave combat leader.[36]

Paratroopers in Action against Armor

The 82nd Airborne Division—employed for the first time as a division—would jump half of its force to assist in the seizure of the airfields in the vicinity of Gela and also hold part of its force as a reserve. The initial field order called for the 505th Parachute Infantry Regiment (PIR) to destroy enemy strongpoints at key road intersections as well as to seize key hills that overlooked these road junctions. Reminiscent of the issues encountered in North Africa, both scheduled jumps struggled to accomplish their assigned tasks due to poor plane navigation and friendly fire from Allied ground and naval forces.[37] However, the little groups of paratroopers managed to cobble together a couple battalion-sized actions that demonstrated dynamic leadership.

Lieutenant Colonel Arthur Gorham, commander of 1st Battalion, 505th Parachute Infantry Regiment, assembled just under a hundred men from his five companies halfway between Niscemi and Piano Lupo. In his first twelve hours on the ground, Gorham led his soldiers to destroy an Italian strongpoint, set up a blocking position, and knocked three enemy vehicles out of action from this position. Gorham quickly realized that he could not hold this position any longer once Axis artillery began to fall on his troops' positions. Also, he did not have the communications necessary to orchestrate naval gunfire against a column of Italian tanks that started to flank his position. Lieutenant Colonel Gorham then moved his force to their assigned objective—the crossroads at Piano Lupo—to make contact with RCT 16 as well as more of his paratroopers.[38] As the men arrived at the crossroads Gorham identified and immediately assaulted a fortified farmhouse and trench lines to seize this objective. The paratroopers eventually overwhelmed the combined sixty-man force of Germans and Italians with only their organic weapons.[39] After this engagement, Gorham located the lead scouts from RCT 16 and passed off a captured map of other enemy positions around the crossroads; the 1st Battalion of the 16th Infantry used it to clear the fighting positions. Gorham then attached his small detachment to the 2nd Battalion, 16th Infantry, of the 1st Infantry Division. In his first day in combat, Arthur Gorham had successfully led his unit in two combat actions and displayed sound decision making in his movement to Piano Lupo. His actions over the next two days against the Axis counterattacks would showcase the courage and calmness under fire required to lead soldiers at the battalion level.

Despite the ferocity of the German armored counterattack, the 1st and 2nd Battalions, 16th Infantry, along with Gorham's band of paratroopers, managed to hold the road junction at Piano Lupo. A mix of about ten Mark III and IV tanks supported by a battalion of infantry assaulted the American infantrymen. While the tanks engaged the defenders, Gorham leaped from position to position, directing rifle and machine-gun fire into the open ports of the tanks.[40] In one instance, he personally manned a bazooka to knock a tank out of action to finally stave off the German counterattack. As the Axis forces consolidated, the 1st Infantry went on the offensive the next day, with Gorham's troops leading the way west toward Niscemi. The pressure of the Big Red One's advance forced the German armored forces to react. As Gorham set his men into position, six Mark VI Tiger tanks supported by armored cars, half-tracks, two platoons of infantry, and artillery from Niscemi attacked the ridgeline above Casa del Priolo. The initial contact almost broke the main line of resistance. Lieutenant Colonel Gorham recognized the grave circumstances and again personally destroyed a tank with a bazooka. He then attempted to destroy another tank with hand grenades and his rifle, but was killed by the main gun of a Tiger tank.[41] His personal actions and ability to inspire his men to hold their positions allowed enough time for RCT 16 to emplace its cannon company and two tank platoons and coordinate for three artillery battalions in support.[42] Gorham's personal courage stands out as an extreme example, but mostly due to the conditions. Faced with combined arms armored formations and few tools to defeat them, Gorham had to use personal example and exposure to the enemy to motivate his men in these circumstances. His aggressive stance along with other key battalion leaders from the 1st Infantry Division facilitated the Big Red One's ability to defeat all the Axis counterattacks against the beachhead.

The action by the 3rd Battalion, 504th PIR, was a more typical event for the paratroopers, but highlights the leadership traits of the battalion commanders. Lieutenant Colonel Charles Kouns, battalion commander for 3rd Battalion, 504th PIR, and attached to the 505th PIR, led a small action of twenty-six soldiers against a battalion-sized task force of the Hermann Goering Division that approached Niscemi. In this small action, Kouns divided his force into two to maneuver against the column. The paratroopers destroyed two vehicles with bazookas and forced the column to delay its movement by two hours. This action was typical for most of the small groups of paratroopers wreaking havoc across the Sicilian countryside.[43]

In addition to the 1st Infantry Division stand outside of Gela, Major Krause's 3rd Battalion, 505th Parachute Infantry Regiment, established a defense along the Biazza Ridge to secure the Gela highway. Krause's 180-man force integrated three 75mm pack howitzers, two 57mm anti-tank guns, five rocket launchers, 155mm artillery, and naval gunfire to defeat a counterattack by a battalion from the Hermann Goering Division. In order to prevent the Mark IV and VI tanks from overrunning the battalion's command post, Krause's indirect-fire plan forced the counterattack to stall. Krause's paratroopers knocked out four tanks, two armored cars, and twelve mortars, and killed or captured over 100

3rd Battalion, 505th PIR, Paratroopers on Biazza Ridge. World War II Signal Corps, MHI, http://www.carlisle.army.mil/AHEC/mediagallery/photo Gallery.cfm?id=50.

Germans at a price of 43 U.S. soldiers killed.[44] The gutsy defense on Biazza Ridge was the last setback that convinced General Conrath to break contact with American forces near Gela.[45]

After the consolidation of the beachhead, the 82nd Airborne Division assisted the 3rd Infantry and 2nd Armored Divisions' race to Palermo. The 82nd Division was pulled out of the line to prepare for the invasion of Italy once Patton got to Palermo. At a cost of just over 1,000 casualties, the paratroopers had advanced the consolidation of the beachhead by approximately two days and fought a critical defense at Biazza Ridge.[46] Usually a little younger than the battalion leadership of the other divisions, these officers—men like Gorham—proved the importance of direct leadership under the most lethal conditions. The photo of the paratroopers on Biazza Ridge tells the tale. These soldiers often had little more than their rifles, grenades, bazookas, and entrenching tools to dig foxholes against mechanized formations. All of the infantrymen from the other divisions were similarly armed. Effective battalion leadership not only had to integrate artillery, close air support (when possible), tanks, and antitank guns, but they had to inspire their men to stay and fight formidable forces.

The final summary of the "Lessons from the Sicilian Campaign" compiled by combat observers and commanders in Sicily emphasized the importance of leadership, prior training, and preparation with this statement:

> Several of the divisions that took part in the assault on Sicily and participated throughout the campaign that followed had had no previous combat experience. These divisions acquitted themselves in excellent style and conducted their operations with a degree of combat efficiency comparable to that of the veteran units of the Tunisian Campaign. The reason for this high degree of success on the part of units not hitherto in action can be attributed only to the excellence of their leadership, prior training, and preparation for combat. . . . A poorly trained unit cannot learn profitably by combat, since it is not prepared to make the most of the battle experience it receives and the confidence that battle experience imparts to the soundly trained organization.[47]

With a solid training base, the battalion leadership on Sicily, with and without combat experience, led their formations effectively and often under the most difficult of conditions on the Licata beaches, the Gela plain, or the heavily defended road network through the Sicilian interior.

Conclusion

By the time the U.S. Army reached Messina, Sicily, it had proven itself as an adaptive, flexible, and lethal military force capable of defeating Axis forces on the Mediterranean battlefields. As the war progressed, the American army gradually rectified its lackluster performance at the operational level of war to match its tactical capabilities.[1] Much of the early success relied on the battalion leadership that executed these early campaigns. The organization of the army for World War II employed an effective system to place its midlevel field-grade officers in key battalion leadership billets. The North African and Sicilian campaigns validated this cadre system as a legitimate practice. These officers succeeded for multiple reasons. Their robust prewar experiences prepared them to execute their profession and developed complementary personal factors to handle the stressors of combat. These officers also maintained command climates conducive to learning by incorporating citizen-soldiers' contributions. Even though none had prior combat experience, their training, maturity, and education facilitated their adaptation to the tactical factors encountered in combat.

The tactical success of these battalion leaders rested on the foundation of their prewar experiences. The officer corps's solid educational experience, mostly from West Point and four-year universities, provided the base for the remainder of their careers. While it is impossible to chronicle the curriculum of the dozens and dozens of universities that produced ROTC graduates, a detailed analysis of a cadet's four years at West Point confirms the rigorous and diverse exposure future officers received during college. The balanced approach to the sciences and the arts prompted the future officers' outlook on warfare to consider all factors. Approximately 85 percent of all the MTO battalion leadership graduated from West Point, and almost 95 percent of the survivors completed twenty-year careers in the army. Innovative instruction and discussion in the classroom encouraged the cadets to pursue a lifetime study of their profession once they left the academy. Since many of the techniques and procedures adopted at West Point also were used at other universities, the ROTC cadets received a similar academic background.

Once the newly minted officers left their schools, they would eventually matriculate to their branch schools about four years after graduation. The year spent at the Infantry and Cavalry Schools continued to promote the mastery of their chosen profession and built on much of the history and military training taught at West Point. Officers went to their branch schools with one or two assignments under their belts to add deeper perspective to their craft. They spent hours reading about both the science and the art of war, attempting to understand the gap between the tactical and operational levels of war. Furthermore, their instructors coached them through tactical vignettes utilizing the most current technology and examples. This experience also forced them to develop techniques and procedures to train the battalion organizations they would later command. A yearlong investment in the professional military education of an officer, the branch schools came at a high cost in time as well as instructors. For an army that had decided to invest in its people more than machines during the budget crunches of the 1930s, this investment made sense if these officers should be needed to serve as a cadre to expand the army.

While at war, the U.S. Army maintained this commitment to the combat effectiveness of its officer corps. General Marshall's stance toward officer education prior to entry into World War II makes this clear. Disagreeing with Secretary of War Henry Stimson, Marshall would not directly commission officers serving in combat-arms positions. He required that all of these officers go through Officer Candidate School or be graduates of West Point to lead men in combat.[2] Once commissioned, other senior leaders believed in the constant instruction of the officers to develop and maintain proficiency. In his Commanding General's Report on Combat Experience of 1st Armored Division, Ernest Harmon emphasized, "More stress should be given to the instruction of junior officers and non-commissioned officers concerning their duties and responsibilities as leaders."[3] In a time of global war, the U.S. Army's leaders placed so much primacy on leader development that they removed proven combat leaders from their billets to place them as instructors in the United States to train future combat leaders. In the late fall of 1944, two captains—John McWatters and Robert Van Zant—left their assignments as company commanders in the 13th Armored Regiment on campaign in Italy to report to Fort Knox, Kentucky, as instructors in the Tactics Department of the Armored School. Both instructors taught classes about joint air-ground operations, combat experience, and employment of a tank battalion with infantry in the attack, to name a few examples.[4] Before they were allowed to instruct, they had to go through numerous "murder boards" and tryouts to become certified.[5] The supervisors and peers graded each instructor on a scale of one to five in the categories of appearance, voice, vocabulary, mannerisms, presence, preparation, discipline, presentation, and training aids. Not coincidentally, one of the certifying instructors was Colonel William Kern, former commander of the 1st Battalion, 6th Armored Infantry Regiment, 1st Armored Division. As in the 1930s, the army continued to invest in its leaders during a time of war even though these combat-tested leaders still could have led units.

In addition to the commissioning source and professional military education of the branch schools, the MTO battalion leadership benefited mightily from their assignments and the professionalism of the U.S. Army prior to World War II. Every one of them participated in unit training that exposed them to individual as well as unit-level exercises up to the division level. Whether experimenting with air-ground coordination in Wyoming, conducting battalion force-on-force maneuvers at Fort Ord, California, or conducting defensive plans in the Philippines or Canal Zone, the young officers became proficient at leading men at the battalion level and below. Many of them participated in the Army General Headquarters maneuvers from 1939 through 1941. Demanding missions over varied terrain, umpires, and the incorporation of all the combat arms created the ultimate peacetime laboratory for the officers. Ultimately, the officers walked away from these exercises with a better understanding of the operational challenges—such as air-ground coordination and maintenance issues over long distances—that they would later encounter in the MTO. The exercises also let them become familiar with the trials of integrating new draftees since the expanding army incorporated them wherever they could during the maneuvers. Lastly, the army's adoption of abbreviated field orders forced the MTO battalion leadership to operate off mission-type orders and develop internal standing operating procedures for their own units. All of these peacetime exercises became the outdoor classroom for the junior officer to see how combat might look. In order to regulate officer performance in the army prior to World War II, the officer corps employed a valuable system of officer efficiency reports that marked successful officers committed to a lifelong calling as an officer.[6] This healthy professional environment established a common set of personal factors that shaped the MTO battalion leadership.

As a population, the battalion leadership exhibited similar personal factors in combat. The battalion level was normally the last organization that an officer could routinely lead with direct leadership. Hence, leadership by example became paramount in combat, especially in dire situations. The battalion leadership of the 1st, 3rd, 9th, 34th, and 45th Infantry Divisions and the 1st Armored Division consistently displayed personal valor. Almost 90 percent of them earned at least one valorous award, and just over 60 percent sustained combat wounds resulting in a Purple Heart.[7] Usually not older than thirty-five, the battalion leadership exercised the appropriate level of energy to lead men on 30-mile road marches, clamber up rocky hillsides in contact, or fight a tank over rough terrain. Unlike regimental commanders, who were much older and commissioned during or immediately following World War I, the MTO battalion leadership had matured as officers in a different era. Regimental leadership could be more hit or miss than the battalion leadership. Some regimental commanders were almost relieved after only one day of combat, such as Colonel Forrest E. Cookson of the 180th Infantry Regiment, 45th Infantry Division. Lieutenant General Patton offered the regiment to Lieutenant Colonel Darby, who refused so he could continue to command the 1st Ranger Battalion. On the other hand, Colonel Mark Boatner Jr., the commander for 168th Infantry Regiment, had changed branches from engineers to infantry when the war started so he could fight. A member of the USMA Class of 1918, he concealed three broken ribs resulting from a North African jeep accident so as to continue as a commander and not be placed on staff. He would later earn a Silver Star during a reconnaissance of the Gustav Line near Mount Cassino, Italy.[8] At the battalion level, there were fewer legacy soldiers than the regimental level and fewer discrepancies between their capabilities.

Their dedication to the profession of arms and discipline formed the basis of their common traits. The focused approach to continually train their battalions and integrate new soldiers prior to new operations stands as a tangible measurement. Both the leaders themselves and the combat observers could not emphasize enough the criticality of realistic training. Every one of the battalion-level leaders had conducted quality training in the "Old Army" as well as the army maneuvers from 1939 to 1941. They viewed training as a starting point to comprehend such concepts as managing tank-infantry coordination or calling for indirect fire so soldiers could execute these tasks in war. The repetitive nature of the training instilled discipline in the battalions as well as their leaders. Battalion leadership confidence improved the more they trained their units, resulting in leaders who could calmly lead their formations in combat. For those who survived the war, their length in service and final rank indicate their obligation to their profession as a U.S. Army officer. Nineteen attained general rank, and approximately seventy of them retired from the army as full colonels. The time in service averaged over twenty-three years.[9] The time in service is more significant than the rank attained, since it indicates the sense of duty these officers maintained even after the war ended.

When they were not reading about military operations or participating in them, the contact with other officers and the CCC expanded their leadership horizons beyond West Point and the army they already knew. Since the majority of their peers hailed from the same alma mater, the interaction with Thomason Act officers and the Civilian Conservation Corps broadened their exposure to other groups that they would work with in the future. The influx of Thomason Act officers gradually eroded any biases that a large number of West Point regulars might have had and allowed them to mentor young officers. Most importantly, the regulars recognized that they could develop junior officers from any commissioning source to be successful. As the army absorbed thousands of OCS officers who would lead platoons and companies in the first Mediterranean campaigns, these new officers were accepted, valued, and integrated as key leaders in the battalions. Citizen-soldiers largely accepted their regular army officer leadership based on the New Deal's effect on American society prior to the war. Some of these experiences, such as the CCC, furthered the regular officers' grasp of leading men in difficult circumstances. Across the entire country, the CCC charged young officers with the hefty responsibilities of managing companies of young men to complete a wide array of projects. These interactions and responsibilities made the MTO battalion leadership better officers.

The soldiers who filled the ranks of the U.S. Army's infantry and armor battalions were not usually the pick of the litter according to the Army General Classification Test (AGCT). The Army Air Corps and Army Service Forces acquired a higher percentage of Class I and II soldiers. Compounding this issue was the sluggish replacement system, which often suffered from mismanagement at replacement depots. While critics can easily point out the system's flaws to degrade the performance of U.S. Army combat formations, the MTO battalion leadership effectively integrated these new soldiers into their units. Their prewar knowledge coupled with aggressive training between combat operations transformed these soldiers into key members of their teams.

The shortened basic training, time wasted at depots, and lack of volunteerism to join ground combat units presented challenges for the battalion leadership to be sure. However, they never blamed "inferior" soldiers for battlefield losses or challenges. When they identified shortcomings, these field-grade officers developed training programs to rectify the issues, and soldiers often contributed to developing improved SOPs. For example, the 1st Armored Division conducted two weeks of training following the Kasserine battles to integrate new soldiers and implement lessons learned. Likewise, the 1st and 3rd Infantry Divisions absorbed thousands of replacements following the Tunisian campaign and executed a two-month training program for HUSKY. Once the Allies came ashore at Normandy, opportunities and time to train certainly became more scarce, and commanders had to make deliberate decisions to withdraw units from combat to train based on their effectiveness or need to address a particularly taxing challenge, such as overcoming the Normandy hedgerow defenses. The Mediterranean theater of operations' more limited strategic objectives and geography permitted leadership to dedicate more time to soldier integration.

The disjointed operational approach to many of the Tunisian operations and both the TORCH and HUSKY landings required several leaders to fight their battalions separately. The port assaults of BRUSHWOOD (Port Lyautey), TERMINAL (Algiers), and RESERVIST (Oran), the Blade Force in the race for Tunisia, any of the airborne operations, and the rangers' actions are a few examples of this independent fighting. In these actions, direct oversight and guidance from a regimental or combat command headquarters did not exist or could not affect the situation based on battlefield conditions. As staffs and commanders improved at the regimental level and above, they improved the command and control of battalions to synchronize movement, firepower, and maneuver. As discussed, many of these battalion-sized actions turned out to be disasters, such as the 509th's Operation VILLAIN, but they usually failed for reasons well outside the control of battalion leadership.

The U.S. Army's adoption of mission-oriented orders prior to World War II armed the MTO battalion leadership with the mind-set to operate with limited supervision from their higher headquarters. Whether on the attack or planning for a deliberate amphibious operation, battalion leadership digested minimalist field orders to prepare their units. They became familiar with this decentralization throughout their training leading up to their deployment to the shores of North Africa. The 3rd Infantry field order for the invasion of Sicily stands out as the best example of this skill set. All of the battalions from RCT 7 adapted to significant battlefield friction on Red Beach to achieve the handful of tasks assigned in the original field order. The battalion leaders understood their higher headquarters' intent to accomplish the operational and tactical goals of Operation HUSKY.

The officers leading the battalions in the MTO had absorbed plenty of history and theory while attending college and the branch schools. Building on this intellectual groundwork, they continually adapted and practiced the application of U.S. Army doctrine on the battlefield. When recorded by combat observers or the battalion leaders, they all concurred that the vast majority of U.S. doctrine was "sound." Every one of them used the adjective "sound." This common description of doctrine showed that all the leaders were familiar with the manuals and found them useful when balanced against history, theory, common sense, and battlefield conditions.[10] The actions of Lieutenant Colonel Hightower at Faid Pass or Lieutenant Colonel York at Longstop Hill or Lieutenant Colonel Crawford near Gela all reveal battalion leadership comfortable with the application of doctrine. In the absence of doctrine or stout operations orders, leaders such as Howze and Cole employed appropriate SOPs developed in training or augmented their formations with additional assets. Light infantry battalions, such as the 3rd Battalion, 505th PIR, in its stand on Biazza Ridge, had to often resort to maximizing whatever firepower they could get their hands on to reduce objectives or fight armored formations. U.S. Army doctrine was one tool that the battalion leadership relied up, but not the only one.

U.S. Army casualty figures in the MTO do not point to inadequate battalion-level leadership that bludgeoned and overwhelmed Axis forces with vast amounts of soldiers, resources, and superior firepower. Numerous factors come together in war to explain casualty rates. However, the depiction of the battle casualties in the three charts indicates that there is no discernible trend of units getting better as the war progressed. Combat effectiveness and casualties sustained could ebb and flow depending on time available to train, missions assigned, or amount of time in combat. For example, the MTO casualties for 1942–1943 chart reveals spikes in casualties when the Allies conducted sustained campaigns and valleys when little to no action occurred, as was the case prior to Operation HUSKY. Also, divisions that remained in the MTO till the end of the war suffered similar casualty rates—the 3rd Infantry Division sustained the most due to its higher days in combat. The campaign comparison offers the best evidence that units did not suffer

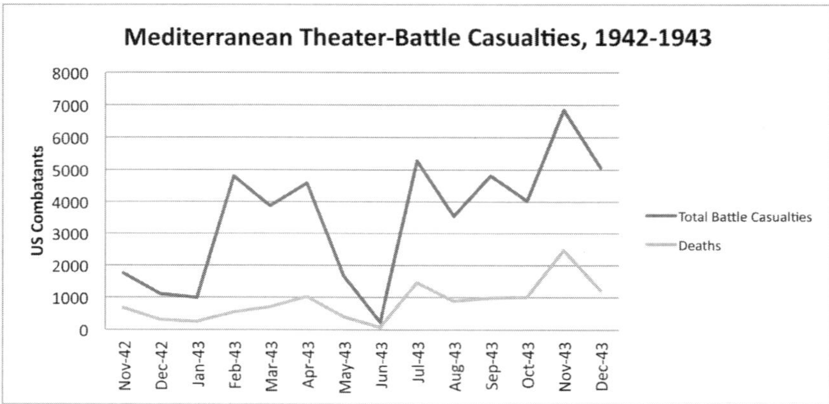

Mediterranean Theater—Battle Casualties, 1942–1943. This chart chronicles all U.S. Army battle deaths for the Mediterranean Theater of Operations to include officers, enlisted, and air corps personnel. Battle Casualties and Nonbattle Deaths in World War II, Final Report, 7 December 1941–31 December 1946, 10 May 1954, Statistical and Accounting Branch Office of the Adjutant General, NARA II, 34.

higher casualty rates, more than units in later campaigns, due to inexperience. The Tunisian campaign often receives the most criticism of unnecessary casualties caused by incompetent leaders, but its casualty rates and seven-month length do not indicate this. The five-month-long Naples-Foggia campaign sustained similar rates, and Anzio, another five-month campaign, sustained almost double the rate of the Tunisian campaign. Both Sicily and Normandy—campaigns of less than two months—reveal much higher rates due to the nature of the combat, terrain, and number of units involved. The dearth of operational experience across the U.S. Army for the North African and Sicilian campaigns could have led to discernibly higher casualty rates for these earlier invasions. The effective battalion leadership during these operations was one factor that mitigated the detrimental effects of an army weak in operational expertise.

The German superiority in tanks often stands out as an explanation for the defeat of American armor on the battlefield. There is no doubt that Germany built some tanks with better armor and more powerful guns. This discrepancy is most evident in the actions against the M3 Stuart tanks, but also against the Grant and Sherman. Production factors and a penchant for mobility dictated that the American tanker

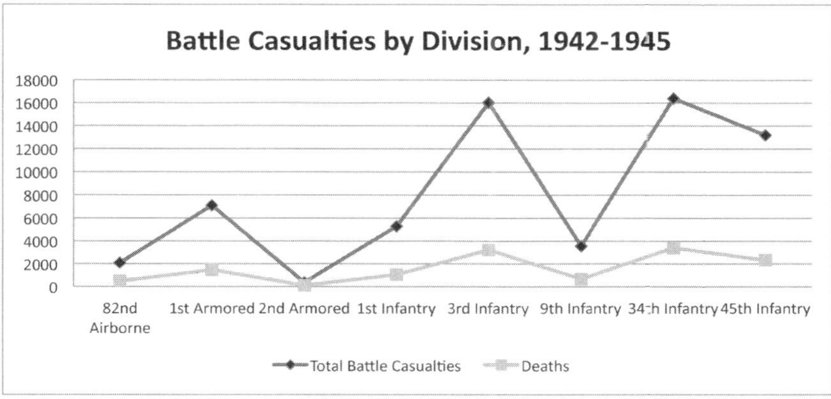

Battle Casualties by Division, 1942–1945. Casualties by division in MTO only. Battle Casualties and Nonbattle Deaths in World War II, Final Report, 7 December 1941–31 December 1946, 10 May 1954, Statistical and Accounting Branch Office of the Adjutant General, NARA II, 86–87.

would ride the Sherman into battle until the very end of the war. Just as leaders adapted doctrine, combat leaders had to adjust for the technological deficiencies and overcome this gap through the savvy use of tactics and combined arms, much as Howze executed near Mateur and the Big Red One's battalions did at Oran. The data in chapter 4 tell the tale: superior ratios, regardless of type of tanks or antitank guns, usually determined the outcome of battles, particularly in the numbers exhibited by the Germans at the opening battles of Kasserine Pass. This assertion holds even more true in North Africa, where 85 percent of tank casualties came from gunfire at average ranges of almost 900 yards. In the MTO, superior German armament did not account alone for U.S. defeats at the battalion level.

Early in the war, the army struggled to leverage numerous functions, especially at the operational level of war, that would become great strengths as World War II progressed. Throughout TORCH and HUSKY, maneuver battalions could never count on consistent or responsive close air support or air superiority. The poor integration of airpower handicapped the timely incorporation of photographic reconnaissance to support ground maneuver. Even though the Allies broke the Enigma code in December 1942, they had not yet developed the appropriate expertise to employ the information gleaned from Ultra for ground warfare. Operational leadership and corps-level staffs initially strug-

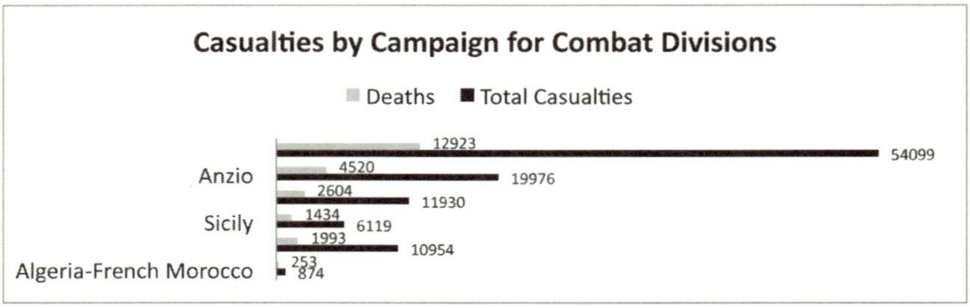

Casualties by Campaign for Combat Divisions. The bar graph compares casualties for five different campaigns. Algeria–French Morocco, Tunisia, and Sicily all relate directly to this study. Naples-Foggia, Anzio, and Normandy were included to provide three comparison points to the first three MTO campaigns. Battle Casualties and Nonbattle Deaths in World War II, Final Report, 7 December 1941–31 December 1946, 10 May 1954, Statistical and Accounting Branch Office of the Adjutant General, NARA II, 92–93.

gled to manage the fluidity of corps-level maneuver in mechanized warfare. These staffs had not yet mastered the ability to utilize wireless communications to its fullest extent and often remained tied to a telephone.[11] At this level, the corps headquarters had to peer deeper into the future than its divisions to synchronize the maneuver of these units with their allies and with air and naval assets. Mostly, corps commanders could not do this yet. The challenges of amphibious operations often captured the attention of corps-level staffs, but not other aspects at the operational level of war. Major General Fredendall's inability to employ corps-level reconnaissance or synchronize all the units at his disposal prior to the Kasserine Pass battles stands out as the most egregious example of these operational failures.[12] Fortunately, the Allies recovered from these shortcomings by studying the lessons learned from the North African and Sicilian campaigns.

After the Battle of El Guettar and while the Allies prepared to expel Axis forces from North Africa for good, General Eisenhower visited the II Corps headquarters. In a letter to General Marshall, he made two interesting observations in regard to combat effectiveness. In a conversation with a combat-experienced British liaison officer, the officer claimed that the 1st Infantry Division was "one of the finest tactical organizations that he had ever seen."[13] Before the Americans even left Tunisia, British observers heaped praise on their tactical capabilities. Ike also informed the chief of staff that the 3rd Infantry Division would "give a good account of itself" as it moved up to the front despite its inexperience because it was a good organization with an outstanding commander.[14] At Ike's level of command, he evaluated divisions even though most of the fighting up to El Guettar was not done by consolidated divisions. Corps commanders such as Patton did the same thing, as evidenced by his diary comments of 13 March: "34th is too defensive. 9th has 'Valor of Ignorance.' 1st is good. 1st Armored is timid."[15] If a division got a bad reputation—such as 1st Armored Division after the Kasserine Pass battles—then that stigma touched the whole division regardless of actions by the battalions within it. Furthermore, the 3rd Infantry Division's paltry combat experience up to that point did not concern Eisenhower about the unit's future success. Quality division commanders such as Allen and Truscott certainly helped to establish effective divisions. However, lack of combat experience did not stop these organizations from defeating a veteran enemy. Another Eisenhower vignette sheds more light on the issue of being blooded.

About two hours before the Axis launched Operation FRUEHLINGSWIND, Eisenhower's driver fell asleep and promptly ran into a ditch. By the time Eisenhower got back to his headquarters, the attack had begun its move westward. Lieutenant Colonels John Waters, John Van Vliet Jr., and Louis Hightower forwarded accurate reports of enemy strength and location up through their chain of command in the naïve hopes that a well-planned counterattack and air support would come to the rescue. Unfortunately, as Ike later pointed out, although "very accurate reports were submitted by the American troops to General Anderson concerning the strength and direction of the German attack through Faid, these reports were discounted by the Army and AFHQ Intelligence divisions as the exaggeration of green, untried troops."[16] These vast generalizations about greenness before, during, and after engagements in North Africa, particularly Kasserine Pass, lack any substance for evaluating battalion leadership, especially of the 1st Armored and 34th Infantry Divisions. While combat experience does educate and harden leaders, that education can come at a high and uneven price. Tried leaders like Lieutenant Colonel John Todd got killed; some, like Waters and Alger, were captured; others were wounded and forced to miss days in the line, like Gardiner. Some officers, such as Howze, were thrown into the jaws of combat at a moment's notice, while others avoided these mishaps and performed well, like Hightower, and others faltered, like Bruss. In fact, one of Hightower's former platoon leaders, who recognized him as one of the best leaders he served under in combat, commented, "Historians tend to blame our defeat on 'Green American Troops' but this was not the case in my judgement. We were beaten because our Division was fed piece meal [*sic*] to the enemy."[17]

In North Africa and Sicily, sound leadership, regardless of combat experience, determined how a unit would stand up in battlefield engagements. This concept can be extrapolated beyond the shores of the Mediterranean. Actual tactical combat experience could not be institutionalized throughout the U.S. Army's eighty-nine divisions. As an organization, the U.S. Army got better operationally as it consolidated lessons learned from TORCH and HUSKY. The pendulum of division, regiment, and battalion combat effectiveness would constantly swing back and forth based on the amount of enemy contact, attrition, missions assigned, and time to train. The battalion leadership of the MTO divisions came mostly from the regular army of 1930s. They possessed the requisite skill set and professionalism to train and lead their soldiers effectively without prior combat experience. These officers could rapidly encourage tactical adaptation during combat operations. Divisions formed later, in late 1942 and 1943, did not have cadres of quality battalion leadership to prepare these units for the rigors of first combat.[18] Throughout the North African and Sicilian campaigns, the U.S. Army entered combat with effective battalion leadership that could achieve tactical success despite the U.S. Army's operational challenges and the rigors of combat against the Axis.

Throughout his time in North Africa and Italy, Hamilton Howze carried a copy of *The Principles of War* by Carl von Clausewitz; he later recollected the importance of this work to him as a combat leader:

> I found the book eminently sustaining in times of stress—it was something to cling to, to refer to when an operational plan went awry, when units went kiting off on the wrong road, when subordinates seemed to lose their minds, or when your luck was bad. In other words, on a normal battle day. The book was reassurance, a friendly voice reminding me of the inescapable, grinding friction that characterizes even well-conceived battle operations, and this if one has planned reasonably well, has made allowances for difficulties and delays, and has taken every possible measure to make success a certainty, things just might come out pretty well.[19]

This insightful comment captures the mind-set exhibited by the battalion leadership throughout its early campaigns in the MTO. As a group, these leaders provided the agile leadership to develop and lead combat effective battalions during the U.S. Army's first critical campaigns despite serious operational handicaps. As a group, they evolved during the 1930s and in combat to produce successful battalion leadership when the Allies needed it most. Their battle-scarred and dirty success is a testament to the necessity of professional armies to remain grounded in quality professional military education and rigorous training in both peacetime and war. This preparation assures the ability to develop large armies and to be ready for combat at the outset of any war a country might face.

Historiography Essay

The sheer scope of World War II has encouraged a hefty production of historical works, especially about the United States' efforts. The specter of the Cold War heavily influenced the analysis of the U.S. Army in World War II in a negative way to both degrade the Soviet *Ostfront* victory and uphold Germany as a worthy ally. The Cold War meltdown in 1989 offered a new course owed mostly to the resurrection of the Red Army's reputation. The fortieth, fiftieth, and sixtieth anniversaries of World War II coupled with the overwhelming popularity of films and books such as *Band of Brothers* and *Saving Private Ryan* encouraged a healthy reevaluation of the U.S. Army.[1] The collapsing Soviet state, veterans' willingness to talk, and commemoration of Allied victory spurred a few dedicated American historians to give more credit to the U.S. Army starting in the 1980s. Despite this resurgent effort, no works had given proper attention and analysis to the field-grade officers who led the battalions at the beginning of the war. Furthermore, scholars have not evaluated how field-grade officers created combat-effective units at the battalion level. So, while the U.S. Army has received its fair share of historical revival, most of the publications have showcased the entire organization, the generalship, or the average soldier.

Historians began questioning the combat effectiveness of the U.S. Army almost as soon as the celebrations of V-E and V-J ended. The first and best known assessments of the U.S. Army came from Colonel S. L. A. "Slam" Marshall's *Men against Fire* and Colonel Trevor Dupuy's *Numbers, Predictions, and War*. Marshall claimed that only 25 percent of U.S. infantrymen actually fired their weapons in combat, while Dupuy used a dizzying array of statistical analysis in his quantified judgment model to conclude that the Wehrmacht was 20 percent more capable than U.S. divisions.[2] Russell Weigley's monumental work, *Eisenhower's Lieutenants*, and his metanarrative of U.S. warfare, *The American Way of War*, popularized Dupuy's assessment. Weigley claimed that the U.S. Army was incapable of escaping its frontier legacy, developing inspired leadership, or building capable maneuver forces that could operate without prodigious amounts of artillery or logistics. Recently, Adrian Lewis's *The American Culture of War* has tried to extend Weigley's thesis beyond World War II. He argues that the rise of airpower, technology, and increased global commitments changed the culture of America's armed forces. He goes on to say that this process began during World War II with a reliance on technological advances and growth of strategic responsibilities.[3] Unfortunately, the U.S. Army's internal studies centered mostly on the performance of weapon systems such as tanks and antitank guns and not on individuals. As an institution, the U.S. Army did not evaluate the infantryman with a rifle as a system. Besides unit organization and manning, the U.S. Army did not systemically study how its field-grade leadership trained and led its battalions in combat.

The international court of historians has been no kinder than the authors above. Martin van Creveld, a Dutch-born Israeli historian, ironically argued that the Wehrmacht was vastly superior to the U.S. Army. His *Fighting Power: German and US Army Performance, 1939–1945* asserted that the German army's cohesion and training outclassed the bureaucratic U.S. Army. Three British historians joined the fray alongside Creveld. Max Hastings's *Overlord* asserts that the U.S. Army soldiers were nothing more than "a nation of civilians in uniform."[4] John Keegan's *Six Armies in Normandy* and John Ellis's *Brute Force* heartily agreed with Weigley and Creveld.[5] Finally, Jörg Muth's *Command Culture* has greatly criticized the U.S. Army's interwar professional military education and commissioning programs.

On the other hand, early American writers—usually participants or embedded journalists in the war—heaped too much praise onto United States servicemen. Ernie Pyle is the epitome of these writers. From the very beginning in North Africa to his untimely demise on Okinawa, Pyle viewed himself as one of the infantrymen fighting the war. He rarely relayed information about tactics or maneuvers. He strove to describe the "ordinary man" fighting for everyone back home. Pyle identified with the soldiers, becoming completely subsumed by the intoxicating experience of war. He equated his evolution with that of the fighting men. Even after the battles of Kasserine Pass, Pyle commented, "I saw them in battle and afterward, and there was nothing wrong with the American soldier. . . . The deeper he got into a fight the more of a fighting man he became."[6] This first draft of history propagated by these journalists established a rosy view of the U.S. Army that does not stand under careful scrutiny.[7]

As David Glantz and Jonathan House led the charge to present a more balanced view of the Soviets, other historians challenged most of the previous work on the U.S. Army.[8] Allan Millett's essay "The United States Armed Forces in the Second World War," published in the third volume of his edited three-volume series, *Military Effectiveness*, concluded that the U.S. Army was an effective fighting force when compared to its allies and adversaries. Supporting Millett's broad conclusions, John Brown's *Draftee Division* and numerous articles questioned Dupuy's conclusions. Brown determined that U.S. units outperformed German ones except for panzer units. These two historians paved the way for a collection of key works.[9]

Three works have proved the most useful in framing my argument: Michael Doubler's *Closing with the Enemy*, Peter Mansoor's *The GI Offensive in Europe*, and Russell Hart's *Clash of Arms: How the Allies Won in Normandy*. All three of these books bring a more balanced view of the U.S. Army and conclude that the U.S. Army was an adaptive, flexible, and lethal military force that ultimately won on the battlefields of Europe. James Carafano's *GI Ingenuity* and James Scott Wheeler's *The Big Red One* are two recent works that have built on these conclusions.[10] Due to the sheer number of units involved and the superb organization of U.S. official documents/records from late 1944 to May 1945, these authors focused on the Normandy campaign until final victory in Europe. *The Big Red One* is the exception since it is a complete unit history. This book will extend the argument of these works to the Mediterranean Theater as well as prewar mobilization. Even though all five historians discuss the MTO, their work is rather thin and there is little prewar discussion.[11] Some more recent work has attempted to link the interwar years with U.S. Army performance.[12] This work builds on the five works mentioned above and addresses issues not discussed by more negative accounts of the interwar army experience.[13]

Complementing this discussion of combat effectiveness in the MTO is the organizational learning debate. Unfortunately, the history of organizational learning about the U.S. Army of World War II remains scarce. Doubler, Carafano, and Hart identify the U.S. Army as a learning institution over several years. The U.S. Army's ultimate success in western Europe stands as the culminating event proving their theses. Detailed works that discuss learning organizations and tactical adaption focus on the U.S. Army well after World War II. James Russell's *Innovation, Transformation, and War* offers a robust analysis of organizational learning and military innovation theory. Russell claims that the education, agility, and horizontal learning of battalion-level officers allowed certain U.S. Army and USMC units to rapidly adapt their organizations to counterinsurgency in Iraq, 2005–2007.[14] Richard Duncan Downie's *Learning from Conflict* analyzes the U.S. military in Vietnam and El Salvador, and Terry Terriff looks at the North Atlantic Treaty Organization (NATO) at the end of the Cold War in *The Sources of Military Change*.[15] The study of battalion leaders and the units they led across North Africa and onto the shores of Sicily as learning organizations that produced rapid tactical adaptation stands as a unique contribution to organizational learning literature. Previous studies usually did not focus at the battalion level or on conventional warfare.

In addition to this robust debate on military effectiveness and organizational learning, many scholars have produced numerous works about some aspect of the MTO. Generally, these works fall into three categories: ground campaign, air/naval campaign, or strategy. To further complicate the issue, many of these works are colored by biases such as the continual degradation of Italy as a worthy ally of the Germans.[16] Most of the ground campaign work focuses primarily on the Commonwealth forces versus the combined German and Italian forces from 1940 until late 1942. These are abundant and range in perspective and quality. For example, laudatory works such as Samuel Mitcham's *Triumphant Fox* paint Rommel as a military genius while others such as David Irving's *Trail of the Fox* are more balanced.[17] From the U.S. perspective, Rick Atkinson's *An Army at Dawn* is the most complete and readable narrative of the U.S. Army in North Africa, although others exist, such as Norman Gelb's *Desperate Venture* or Kelly Orr's *Meeting the Fox*. George Howe's excellent volume in the U.S. Army's official history series, *Northwest Africa: Seizing the Initiative in the West*, provides the official U.S. Army interpretation of the campaign. Some works, such as Martin Blumenson's *Kasserine Pass* and his chapter in *America's First Battles*, home in on a specific battle. Most of these works are combat narratives and do not seek to argue combat effectiveness, officer development, or the complexity of combined arms doctrine. However, when these scholars do take a stand, Kasserine Pass figures prominently as a turning point for the U.S. Army's development because of the embarrassing defeat. Many of the horror stories from Kasserine Pass highlight the panic exhibited by U.S. soldiers and their abandonment of equipment. The discipline and training of U.S. soldiers is not a major factor in this volume, but the analysis here has addressed how battalion leaders rectified these shortcomings. This work also brings more objective analysis to the entire campaign and reevaluates conclusions about Kasserine Pass and the officer corps that defeated the Axis forces in Tunisia.[18]

I. S. O. Playfair's five-volume *The Mediterranean and the Middle East* is an exhaustive work on the entire theater; when coupled with books that include Ultra decryptions such as Ralph Bennett's *Ultra and the Mediterranean Strategy*, the volumes provide a sufficient overview. Since this study is about a campaign army, not too much naval history is applicable besides the amphibious landings and logistic support. However, strategic studies such as Michael Howard's *The Mediterranean Strategy* and Carlo D'Este's *World War II in the Mediterranean* provide the essential framework to explain the importance or lack thereof of the U.S. Army's campaigns in North Africa and Sicily.[19] These works form the traditional narrative of the Allies in the Mediterranean. In order to put these in proper context, this book incorporates a healthy compilation of after action reports, operations orders, award narratives, and interviews to bring a fresh and less diluted scrutiny of battalion-level leadership in the MTO.

The prewar/Pacific frontier army is the last significant category of historiography. Generally these works use the Spanish-American War or the closure of the western frontier as their beginning and 1939–1941 as the end date for their studies. Brian Linn's exhaustive *Guardians of Empire* is the archetypical example of this type of work; it ably explains the trials and tribulations of America's Pacific Army. More recently, Edward M. "Mac" Coffman has produced an insightful study of what the regular army was actually doing in the first four decades of the 20th century. In his *The Regulars*, Coffman concludes his work with an anecdote about Winston Churchill, who expressed amazement at how rapidly the U.S. Army expanded with such able leadership, calling it a "mystery as yet unexplained."[20] Coffman offers up a simple explanation for the former prime minister's conundrum. "Virtually all the senior commanders and key staff officers had spent their adult lives in the Regular Army. . . . Over the years of working together, they measured each other's character, dedication, and ability. The round of assignments with increasing responsibilities helped shape them."[21] *The Regulars* offers plenty about well-known officers such as Dwight Eisenhower, Lucian Truscott, Joe Collins, and Lesley McNair—generally division commanders and higher—but there exists a void about the younger officers. How did the regular army shape the junior field-grade officers prior to World War II? With little depth in previous works on this subject, this work aims to develop an understanding of these young officers in their very influential years of development and maturation from 1930 until landing on the shores of North Africa and Sicily.[22]

Notes

Introduction

1. The three generals are Lieutenant General Simon Bolivar Buckner (killed on Okinawa leading the Tenth Army), Major General Maurice Rose (killed leading the 3rd Armored Division closing the Ruhr Pocket), and Brigadier General James Wharton (killed leading the 28th Infantry Division in France).

2. T. Michael Booth and Duncan Spencer, "The Airborne's Watery Triumph," in *No End Save Victory: Perspectives on World War II*, ed. Robert Cowley (New York: G. P. Putnam's Sons, 2001), 441.

3. John S. Brown, "Winning Teams: Mobilization-Related Correlates of Success in American World War II Infantry Divisions" (master's thesis, U.S. Army Command and General Staff College, 1985), 36–37.

4. James Scott Wheeler, *The Big Red One: America's Legendary 1st Infantry Division from World War I to Desert Storm*, Modern War Studies (Lawrence: University Press of Kansas, 2007), 132.

5. Kent Roberts Greenfield, Robert R. Palmer, and Bell I. Wiley, *The Organization of Ground Combat Troops*, United States Army in World War II: Army Ground Forces vol. 2-1 (Washington, D.C.: Center of Military History, United States Army, 1947), 89–90.

6. Ibid., 201–202.

7. Robert R. Palmer, Bell I. Wiley, and William R. Keast, *The Procurement and Training of Ground Combat Troops: The Army Ground Forces*, CMH Pub. vol. 2-2 (Washington, D.C.: Center of Military History, U. S. Army, 2003), 95–97. Unfortunately, there are no Army Ground Forces (AGF) studies on combat arms officers. The U.S. Army produced thirty-eight AGF studies covering a wide array of topics, but nothing on these regulars. Nothing in the thorough "Green" book series addresses this issue either. Throughout the U.S. Army's official history, discussion of the regular officers' procurement and training is rather sparse.

8. Brown, "Winning Teams," 172.

9. Edward M. Coffman and Peter F. Herrly, "The American Regular Army Officer Corps between the World Wars," *Armed Forces and Society* 4, no. 1 (Fall 1977): 62–66.

10. *War Department Annual Reports, 1929–1938*, USMA Special Collections. See the historical data section, usually appendix C.

11. Coffman and Herrly, "Between the World Wars," 56–57. The "Hump" is defined as the 5,229 post-1917 commissioned officers who suffered slow promotion rates. Congress did not approve funds for their promotion beyond their World War I rank. Most of these officers remained lieutenants for almost seventeen years until a 1935 congressional act promoted them to captain. Since they never rose above company-grade rank, these officers spent the majority of their careers in line units.

12. Christopher Gabel, *The U.S. Army GHQ Maneuvers of 1941* (Washington, D.C.: U.S. Government Printing Office, 1981), 197–202.

13. Elizabeth Borgwardt, *A New Deal for the World: America's Vision for Human Rights* (Cambridge, Mass.: Belknap Press of Harvard University Press, 2005), 77–78.

14. General Board, United States Forces, European Theater, "Organization, Equipment, and Tactical Employment of the Armored Division," Study Number 48, 7 November 1945, Center of Military History, Washington, D.C., chapter 3, appendixes 1, 2, and 3.

15. Ibid., "Organization, Equipment, and Tactical Employment of the Infantry Division," Study Number 15, 20 November 1945, section 2 and appendix 2.

16. Ibid., "Organization, Equipment, and Tactical Employment of the Airborne Division," Study Number 16, 20 November 1945, chapter 1, section 3 and appendix 4.

17. Lieutenant Colonel Arthur Gorham is one example. A 1938 USMA graduate and 1/505th PIR battalion commander, Gorham left almost no record of his combat and army experience besides Distinguished Service Cross citations. He was killed in Sicily in July 1943. His own son could compile only a two-page biography about his father. At least, as a result of their deeds, men such as Gorham left some record. Some battalion leaders left almost no record.

18. Paul Kennedy, "History from the Middle: The Case of the Second World War," *Journal of Military History* 74 (January 2010): 35–51. Paul Kennedy coined this term and called for a deeper exploration of "the mid-level managers of war" in this key article.

19. Please see the historiographical essay at the end of this book for a detailed discussion of the relevant World War II scholarship.

20. Michael Doubler, *Closing with the Enemy: How GIs Fought the War in Europe, 1944–1945*, Modern War Studies (Lawrence: University Press of Kansas, 1994), 52–65.

21. Ernest Harmon, with Milton MacKaye, and William Ross MacKaye, *Combat Commander: Autobiography of a Soldier* (Englewood Cliffs, N.J.: Prentice-Hall, 1970), 141.

Chapter 1: West Point Evolution

1. *Official Register of the Officers and Cadets*, United States Military Academy, 30 June 1933, United States Military Academy Special Collections (hereafter referred to as USMA), West Point, New York, 22.

2. Michael King, *William Orlando Darby: A Military Biography* (Hamden, Conn.: Archon Book, 1981), 18–26.

3. Officers Qualification Card, Colonel William O. Darby, William O. Darby Papers, USMA; *The Register of Graduates and Former Cadets of the United States Military Academy, West Point, New York*, 2008, 3–25.

4. Williamson Murray and Allan R. Millett, *A War To Be Won: Fighting the Second World War* (Cambridge, Mass.: Belknap Press of Harvard University Press, 2000), 185.

5. "Rigid Army Methods Gone: West Point Director Says a New Discipline Has Come In," *New York Times*, 24 December 1920.

6. Annual Report of the Superintendent, United States Military Academy, 1920, USMA, 4.

7. Theodore J. Crackel, *West Point: A Bicentennial History* (Lawrence: University Press of Kansas, 2002), 193–194.

8. Annual Reports of the Superintendent, United States Military Academy, 1918–1940, USMA. To assess the academy's growth, I have compared twenty-two annual reports. Often, these reports are merely a short regurgitation of cadet graduation data and budget concerns. However, MacArthur used his first report to shape his vision. Other superintendents provided key information, but none used the reports quite like MacArthur did.

9. Annual Report of the Superintendent, USMA, 1930, USMA, 1.

10. Annual Report of the Superintendent, USMA, 1931, USMA, 1–2.

11. Crackel, *West Point*, 208–209. Crackel rightly identifies the U.S. entrance into World War II as the supreme catalyst for USMA to secure additional training land and expand the training program. The war allowed for more money and resources, but not a substantial shift in the curriculum. The academy still included riding as a major part of the curriculum even in 1943.

12. War Department Annual Report (hereafter referred to as WDAR), USMA, 1932, 1–2. The title and format of this annual report varies slightly throughout the 1930s, but generally includes the same information.

13. Ibid., 2.

14. "Army Increased by 28,000 in Year," *New York Times*, 26 July 1936.

15. Annual Report of the Superintendent, 1929, USMA, 2.

16. Annual Report of the Superintendent, 1931, USMA, 4. The academy had sent language instructors to Spain and France before, but the program expanded to the whole department at this time.

17. Annual Report of the Superintendent, 1933, USMA, 3–4.

18. "Change West Point, Dr. Eliot Advises," *New York Times*, 9 May 1920;

"West Point Education: A Reply to Criticisms by Ex-President Eliot of Harvard," *New York Times*, 10 May 1920.

19. "Information Relative to the Appointment and Admission of Cadets to the United States Military Academy," 1936, in *Official Register of the Officers and Cadets*, United States Military Academy, 30 June 1936, USMA, 147.

20. Annual Report of the Superintendent, 1932 and 1935, USMA.

21. Change in Method of Teaching Civil Engineering and Military History, 11 August 1939, Department of Civil and Military Engineering, Box 3, Series 70, USMA, RG 404, National Archives. All of the USMA records filed into a series fall under RG 404. They are all kept in the USMA Special Collections, 4th floor of the old library, but they are part of the interconnected NARA system.

22. Annual Report of the Superintendent, 1937, USMA, 4–5.

23. "Course in Department of Civil and Military Engineering," Memorandum to Superintendent, 24 January 1940, Department of Civil and Military Engineering, Box 3, Series 70, USMA.

24. "Course in Civil and Military Engineering," 1920–1921, through "Course in Civil and Military Engineering," 1928–1929, Department of Civil and Military Engineering, Box 1, Series 70, USMA.

25. Annual Report of the Superintendent, 1920, USMA, 5.

26. "Course in Department of Civil and Military Engineering," Memorandum to Superintendent, 24 January 1940, Department of Civil and Military Engineering, Box 3, Series 70, USMA. Cadets took eighty-five periods of one hour and twenty minutes in military history. History instructors used four primary textbooks: *Great Captains* by Theodore A. Dodge, *Life of Napoleon* by Antoine Henri Jomini, *American Campaigns* by Matthew Steele, and *Military History of the World War* by Colonel Girard McEntee.

27. "The Sino-Japanese Conflict," and "The Civil War in Spain," Department of Civil and Military Engineering, course in Military History, April 1939, Department of Civil and Military Engineering, Box 2, Series 70, USMA.

28. *Military Review*, all issues 1936–1940, Combined Arms Research Library (hereafter referred to as CARL), U.S. Army Combined Arms Center, Fort Leavenworth, Kansas.

29. "Signal Communications for Third Class," Memorandum, 13 August 1937, Training Schedules, 1926–1940, Department of Tactics, Series 111, USMA.

30. Minor Tactics, Series 111, Training Schedules, 1926–1940, Department of Tactics, USMA, 444–460.

31. Memorandum No. 67, "First Class Trip to Mitchell Field, Long Island," 6 June 1936, and "Notes on Flying of First Class at Mitchell Field, June–July, 1936," Series 111, Training Schedules, Department of Tactics, USMA.

32. While the scorekeeping was all in good fun in this case, it is interesting to note that the interwar army held annual machine-gun competitions that officers and soldiers took very seriously. The competitive spirit thrived in training as well as athletics.

33. "Report of Visit of 1st Class to Posts in Third Corps Area," Memorandum to the Commandant of Cadets, 7 July 1931, Series 111, Training Schedules, Department of Tactics, USMA.

34. "Annex No. 1 to Accompany Operations No. 1," Air Corps Demonstration Unit, Fort Benning, Georgia, n.d., Series 111, Training Schedules, Department of Tactics, USMA.

35. "West Point Class Is Off for Georgia," *New York Times*, 29 July 1937.

36. Sheet No. 1, "Detailed Schedule of Instruction for Period Beginning Wednesday, 5 August to Saturday, 15 August, 1936," Infantry School, Fort Benning, Georgia, First Class, United States Military Academy, n.d., Series 111, Training Schedules, Department of Tactics, USMA Special Collections.

37. Arthur S. Collins Jr., Senior Officer Oral History Program (hereafter referred to as SOOHP), Military History Institute (hereafter referred to as MHI), Ridgway Hall, U.S. Army War College, Carlisle Barracks, Pennsylvania, 39–42.

38. Annual Report of the Superintendent, 1937, USMA, 6–7.

39. "Tactics and Technique of Cavalry," Memorandum from LTC John B. Thompson to Commandant of Cadets, Series 111, Training Schedules, 1926–1940, Department of Tactics, USMA, 1–28.

40. Allotment of Time for Military Instruction Chart, September 1935 to June 1936, Series 111, USMA.

41. Chapter 3, "Winter Training, Second Class," chapter 4, "Winter Training, Third Class," chapter 5, "Winter Training, Fourth Class," Schedules of Instruction, 1932–1938, Series 111, Training Schedules, USMA.

42. Terrain Exercise #1, First Class Tactics, "Rifle Company with Machine Guns in Attack," 1935, Series 111, USMA.

43. "Narrative of Command Post Demonstration," and "Command Post Exercise Demonstration," Schedule of Instruction, 1934–1935, First Class, Series 111, USMA.

44. Arthur S. Collins Jr., SOOHP, MHI, 40–41.

45. Some superintendents even highlighted gains in weight, height, and chest size in their annual reports to mark physical progress. See *Annual Report of the Superintendent*, United States Military Academy, 1932, USMA Special Collections, 4, for a representative example.

46. "Final Results of Fall Intramural Athletics," Memorandum 168, 17 November 1936, Series 111, Training Schedules, 1926–1940, Department of Tactics, USMA Special Collections.

47. "Spring Intramural Athletics," Memorandum No. 37, 13 April 1938, Series 111, Training Schedules, 1926–1940, Department of Tactics, USMA Special Collections.

48. Twenty-one senior officer interviews mention the academy's intramurals specifically, and every officer spoke very highly of their importance. See Hamilton Howze, SOOHP, MHI, 21, for one example.

49. Robert W. Porter Jr., SOOHP, MHI, 65.

50. "Alger, J. D." Series 121, Physical Fitness Cards, 1925–1946, USMA; "Training Schedules," 1930–1938, Series 111, Training Schedules, 1926–1940, USMA. These physical fitness cards are organized alphabetically by year and are the only significant records of the office of physical education that remain at USMA.

51. Steven W. Pope, "An Army of Athletes: Played Fields, Battlefields, and the American Military Sporting Experience, 1890–1920," *Journal of Military History* 59, no. 3 (July 1995): 435–456.

52. G. H. Paine, "Polo in the Philippines," *Infantry Journal* 30 (April 1927): 417–418. Interestingly, this entire volume deals with life in the Philippines in every aspect.

53. Letter from Lawrence R. Dewey Jr. to Steven Barry, 12 August 2008.

54. Richard J. Sothern, "Athletics in the Philippines," *Infantry Journal* 30 (April 1927): 393–394.

55. Hamilton Howze, SOOHP, MHI, Tape 1: 20, 33; Tape 2: 17–18, 23–24; Charles F. Ivins, "The Monkeys Have No Tails in Zamboanga," Box 1, Charles Ivins Papers, MHI, 31–33.

56. The subject of the film *The Long Gray Line*, based on his autobiography *Bringing Up the Brass*, Martin Maher was a rarity. He served for fifty-five years at West Point as both an NCO and civilian swimming instructor. Most instructors rotated every three to five years.

57. Unfortunately, no peer evaluations exist of the cadets. The closest thing approaching a peer evaluation were Howitzer yearbook paragraphs written for each cadet. Traditionally, a fellow cadet—usually a friend or roommate—wrote a brief paragraph that described the cadet's personality and hobbies.

58. Robert Hewitt, SOOHP, MHI, 38.

59. Arthur S. Collins Jr., SOOHP, MHI, 42–43.

60. *Official Register of the Officers and Cadets*, United States Military Academy, 1930–1940, USMA.

61. Arthur S. Collins Jr., SOOHP, MHI, 42–43; John K. Waters, SOOHP, MHI, 34.

62. *Official Register of the Officers and Cadets*, United States Military Academy, 1930, 1934, and 1938, USMA.

63. Arthur S. Collins, SOOHP, 42.

64. Robert W. Porter, SOOHP, MHI, 31; Lewis Sorley, *Thunderbolt: General Creighton Abrams and the Army of His Times* (New York: Simon and Schuster, 1992), 17–19.

65. John K. Waters, SOOHP, MHI, 123.

66. Letter from Drew Boatner to Steven Barry, 5 September 2008.

67. John K. Waters, SOOHP, MHI, 116–127.

68. *Official Register of the Officers and Cadets*, United States Military Academy, 1938, 22–23.

69. Hamilton Howze, SOOHP, MHI, 19.

Chapter 2: Forging Leaders beyond the Hudson

1. Letter from Floyd Felice Forte to Major John W. Merrill, 13 April 1942, Floyd Forte Papers, USMA, 2.

2. Letter from Forte to Merrill, Floyd Forte Papers, USMA, 2; The words to the West Point's "The Alma Mater" (P. S. Reinecke, USMA class of 1911) are: "Hail, Alma Mater dear, to us be ever near, help us thy motto bear through all the years. Let duty be well performed, honor be e'er untarned, country be every armed, West Point, by thee. Guide us, thy sons [recently changed to 'thine own'] aright, teach us by day, by night, to keep thine honor bright, for thee to fight. When we depart from thee, serving on land or sea, May we still loyal be, West Point to thee. And when our work is done, our course on earth is run, may it be said, 'Well done; be thou at peace.' E'er may that line of gray increase from day to day, live, serve, and die, we pray, West Point, for thee."

3. "Military History of Floyd Felice Forte," 18 August 1944, Adjutant General's Office, War Department, Floyd Forte Papers, USMA.

4. *Official Register of the Officers and Cadets*, United States Military Academy, 1934, USMA, 30.

5. Medal of Honor Citation, Second Lieutenant Alexander Nininger Jr., n.d., Alexander Nininger Jr. Papers; H. K. Johnson, "Defense along the Abucay Line," *Military Review* 28, no. 11 (February 1949): 43–53.

6. John E. Olson, "USMA in the Philippines, 1941–42," *Assembly*, July 1993, 11–17; Louis Morton, *The Fall of the Philippines* (Washington, D.C.: U.S. Government Printing Office, 1953), 21–24.

7. Paul D. Harkins, SOOHP, MHI, 12.

8. Arthur S. Collins Jr., SOOHP, MHI, no pages listed. Unfortunately, this interview does not have any page numbers listed, although they are on the table of contents. Making my best guess, I estimate pages 52–54. The information is on tapes C140 and C141.

9. John D. Kenderdine, "Teach Them on the Ground," *Infantry Journal* 43 (May–June 1936): 223–224; Arthur S. Collins Jr., SOOHP, MHI, 50–56 (estimate); Water-borne Target Firing, Memorandum, 19 December 1938, Adjutant General's Office Files, 1926–1929, File 353.17, Hawaiian Department, RG 407, NARA II.

10. Letter from Grethen Heltzel to Steven Barry, 7 February 2009.

11. Hamilton Howze, SOOHP, MHI, Tape 2: 4–6.

12. James M. Gavin interview by Edward Coffman, 15 June 1978, Edward Coffman Papers, George C. Marshall Foundation, Lexington, Virginia, 37.

13. Jean E. Engler, SOOHP, MHI, Tape 1: 43–48.

14. Report of Operations, 13th Cavalry, Fort Riley Exercises, 3 October to 5 November, 1938, I. D. White Papers, Kreitzberg Library, Norwich University Archives and Special Collections, Northfield, Vermont, 1–11; 21–59; 82–85.

15. Cavalry School, United States Army, *Cavalry Combat* (Harrisburg, Pa.:

Telegraph Press, 1937); Jean E. Engler, SOOHP, MHI, Tape 1: 35–37; 45–48; Arthur S. Collins Jr., SOOHP, MHI, 52–55.

16. Cavalry School, United States Army, *Cavalry Combat* (Harrisburg, Pa.: Telegraph Press, 1937); Jean E. Engler, SOOHP, MHI, Tape 1: 35–37; 45–48; Arthur S. Collins Jr., SOOHP, MHI, 52–55.

17. Maurice Rose, "Panama's Irregular Cavalry," *Cavalry Journal*, July–August 1935, 26.

18. Steven L. Ossad and Don R. Marsh, *Major General Maurice Rose: World War II's Greatest Forgotten Commander* (New York: Taylor Trade Publishing, 2003), 101–104.

19. James M. Gavin interview by Edward Coffman, 15 June 1978, Edward Coffman Papers, George C. Marshall Foundation, Lexington, Virginia, 5–6.

20. I have deliberately left out a discussion of the 15th Infantry Regiment in China for three reasons. First, the garrison was closed after the Japanese offensives in 1937. Second, the small size of the garrison and early closure prevented a large number of officers from serving in China. Finally, the 15th did not really conduct any maneuvers, given the limited maneuver space. For a detailed discussion of service in China, see both Dennis L. Noble, *The Eagle and the Dragon: The United States Military in China, 1901–1937* (New York: Greenwood Press, 1990), chapter 3, and Roy K. Flint, "The United States Army on the Pacific Frontier, 1899–1939," in *The American Military and the Far East: Proceedings of the Ninth Military History Symposium, United States Air Force Academy, 1–3 October 1980*, ed. Joe C. Dixon (Colorado Springs, Colo.: United States Air Force Academy and Office of Air Force History Headquarters, USAF, 1980), 139–159.

21. Douglas V. Johnson, "Keep 'em Rollin," unpublished manuscript, MHI, 132.

22. Some scholars have argued the opposite about this routine tactical training. They have concluded that the training did not prepare officers very well for modern warfare. However, their analysis has remained focused on the strategic level of war and does not address the rigors of the small-unit maneuvers. See Russell F. Weigley, *History of the United States Army* (New York: Macmillan, 1967), 402–403; Brian M. Linn, "The Long Twilight of the Frontier Army," *Western Historical Quarterly* 27 (Summer 1996): 147, 152–154, 166; Brian Linn, *Guardians of Empire: The U.S. Army and the Pacific, 1902–1940* (Chapel Hill: University of North Carolina Press, 1997), chapter 10.

23. Johnson, "Keep 'em Rollin," MHI, 138, 148–149, 152–155.

24. Letter to E. M. "Mac" Coffman from BG E. V. Sutherland (USMA 1936), 21 July 1990.

25. Edward S. Miller, *War Plan ORANGE: The U.S. Strategy to Defeat Japan, 1897–1945* (Annapolis, Md.: Naval Institute Press, 1991), 152–155.

26. Letter to E. M. "Mac" Coffman from BG E. V. Sutherland (USMA 1936), 21 July 1990; Edward M. Coffman, *The Regulars: The American Army,*

1898–1941 (Cambridge, Mass.: Belknap Press of Harvard University Press, 2004), 341–342.

27. Annual Report, Fiscal Year 1932, 16 July 1932, Adjutant General's Office File, 1926–1939, File 319.12, Hawaiian Department, RG 407, NARA II, 2–3.

28. Equipment for Machine Gun Waterborne Target Firing at Pokai Bay, T.H., 2 February 1938, Adjutant General's Office Central Files, 1926–1939, File 353.17, Hawaiian Department, RG 407, NARA II, 1–2.

29. Annual Report, Fiscal Year, 1931, Adjutant General's Office Central File, 1926–1939, File 319.12, Hawaiian Department, RG 407, NARA II, 5–8.

30. Report of Formal Tactical Inspection, Hawaiian Department, 1933, 12 December 1933, Adjutant General's Office Central File, 1926–1939, File 333-3, Hawaiian Department, RG 407, NARA II, 5. There are multiple inspection reports for every year.

31. Report of the Commanding Officer, 2d Battalion, 30th Infantry, "Joint Overseas Expeditions: Studies in Connection with the Training, Embarking, and the Landing on a Hostile Shore of a Combat Team in a Joint Overseas Expedition," Command and General Staff School, Fort Leavenworth, Kansas, 1933, 14.

32. John K. Waters, SOOHP, MHI, 63–66.

33. Hamilton Howze, SOOHP, MHI, Tape 2: 1–4.

34. WDAR, 1937, USMA, 33; Gerald M. Brennan, "Military's Peacetime Role: Implications of the Civilian Conservation Corps Experience" (master's thesis, U.S. Army Command and General Staff College, 1986), 93–103; Coffman, *The Regulars*, 242–245.

35. Charles William Johnson, "The Civilian Conservation Corps: The Role of the Army" (Ph.D. diss., University of Michigan, 1968), 68; Robert Hewitt, SOOHP, MHI, 51–54; John K. Waters, SOOHP, MHI, 65–67.

36. John Wayne Garbutt, "We Were Soil Soldiers: The Civilian Conservation Corps Camps of the Lufkin District" (master's thesis, Stephen F. Austin State University, 2007), 32–36; Kenneth Wayne Baldridge, "Nine Years of Achievement: The Civilian Conservation Corps in Utah" (Ph.D. diss., Brigham Young University, 1971), 133–151.

37. Johnson, "Keep 'em Rollin," MHI, 243–248.

38. Ibid., 247–248.

39. Ibid., 259.

40. Ibid., 249–259.

41. Elizabeth Borgwardt, *A New Deal for the World: America's Vision for Human Rights* (Cambridge, Mass.: Belknap Press of Harvard University Press, 2005), 77–78.

42. Ibid., 79.

43. The West Point Academic Board developed the examination. Annual Report of the Superintendent, 1937, USMA, 3.

44. "To Amend the National Defense Act," Report #1241, 74th Congress,

1st Session, Congressional Serial Set; "To Strengthen the Air Corps and to Train Reserve Officers," Report #275, 74th Congress, 1st Session, Congressional Serial Set; "Thomason Act Operation Awaits Necessary Funds," *Washington Post*, 5 April 1936.

45. WDAR, 1936 and 1937, USMA. In the same year that Congress passed the Thomason Act, they also authorized a 600-man increase to cadet strength at West Point. This eventually added 150 cadets to each class. West Point had to implement this increase slowly over a four-year time span with the entrance of each new class.

46. "Training Plan Announced for Reserve Officers," *Washington Post*, 3 May 1936.

47. WDAR, 1937, USMA, 3–4.

48. Arthur S. Collins Jr., SOOHP, MHI, 63–64 (estimate).

49. Samuel A. Montague oral history interview, Truman Library, Independence, Missouri, 8–10; Melvin Zais, SOOHP, MHI, Tape 4: 12–17; Coffman, *The Regulars*, 244–245.

50. See David E. Johnson, *Fast Tanks and Heavy Bombers: Innovation in the U.S. Army, 1917–1945*, Cornell Studies in Security Affairs (Ithaca, N.Y.: Cornell University Press, 1998), 218–229, for a succinct explanation.

51. Leslie Anders, "The Watershed: Forrest Harding's Infantry Journal, 1934–1938," *Military Affairs* 40, no. 1 (February 1976): 12–13; Leslie Anders, "Four Decades of American Military Journalism," *Military Affairs* 41, no. 2 (April 1977): 62; *Infantry School Mailing List* 15 (1938), USMA. The academic department of the Infantry School compiled the *Mailing List* from lesson plans and articles written for professional journals such as the *Infantry Journal*.

52. See *Infantry School Mailing List*, 1934–1939, USMA. Before the 1880s, West Point graduates never attended any professional military education schools about tactics, operations, or strategy. William T. Sherman recognized this shortcoming after the Civil War and began the process of developing an officer education program.

53. *The Infantry School Mailing List, Volume XIII* (Fort Benning, Ga.: The Book Shop), January 1937, USMA, 183–272. Unfortunately, none of the articles list individual authors. However, the front matter of each issue lists the ten to twelve officers who authored the entire work. The faculty collectively developed these articles first as lesson plans.

54. *Cavalry School Mailing List*, 15 December 1933, CARL; see the four issues of *Military Review*, 1933, for a representative listing of the cavalry topics.

55. John K. Waters, SOOHP, MHI, 91–98. Interestingly, the horse tradition still permeates the army to this day, especially at Fort Leavenworth and Fort Riley. Fort Leavenworth still conducts an annual "fox" hunt and maintains significant horse stables and show rings.

56. Cavalry School, *Cavalry Combat*, 506.

57. George F. Hofmann, *Through Mobility We Conquer: The Mechanization of U.S. Cavalry* (Lexington: University Press of Kentucky, 2006), 201–202, 239,

330–332; Cavalry School, *Cavalry Combat*. Instructors from the Cavalry School as well as European officers compiled *Cavalry Combat*, a book of case studies. Throughout the book, the editor appropriately highlighted key doctrine especially as it applied to mechanization. The major criticism of this book is the belief that the horse will have any role on the future battlefield. The book does recognize the diminished importance of the horse, but not its disappearance.

58. Hofmann, *Through Mobility We Conquer*, 330–332.

59. The only shortcoming of the CPXs was the intermittent incorporation of wireless communications.

60. *Training Management*, Academic Division, Cavalry School, 1934, 57–61.

61. Ibid., 31–56.

62. James M. Gavin interview by Edward Coffman, 15 June 1978, Edward Coffman Papers, George C. Marshall Foundation, Lexington, Virginia, 20–22.

63. Frank E. Emery Jr., "What Should Be the Plan of Education for Officers of the Army, Including a Discussion of Individual Study, the Special Service Schools, the Command and General Staff School, and the Army War College," Command and General Staff School, 1930, CARL, 10.

64. Ibid., 8–10.

65. A standard basic officer course for each branch did not become mandatory for second lieutenants until after the Vietnam War.

66. Emery, "What Should be the Plan of Education for Officers of the Army," 14–15. Emery went so far as to advocate that units should maintain copies of every book on the War Department reading lists.

67. Christopher Gabel, *The U.S. Army GHQ Maneuvers of 1941* (Washington, D.C.: U.S. Government Printing Office, 1991), 9–13.

68. "Third Army Maneuvers, May 5–25, 1940," CARL, 1.

69. Ibid., "General Instructions for Control of Corps and Army Maneuvers."

70. 4th Corps Provisional Observation Group, 1940, "Photos of Training Areas in Texas, Louisiana, and Georgia," 187 photos and 23 maps, CARL.

71. "Report on Second and Third Army Maneuvers, 1941," n.d., CARL, 23.

72. "Final Report, Third Army Maneuvers, May 1940, Headquarters," CARL, 37–38, 50–52.

73. Ibid., 47–49.

74. "Report on Second and Third Army Maneuvers, 1941," n.d., CARL, 14–15, 17, 23–24, 26, 35, 37–38, 41. John K. Waters, SOOHP, MHI, 63–65.

75. "US Army Ready to Test Its New War Machine," *Washington Post*, 31 August 1941. Both the Louisiana and the Carolina maneuvers had approximately 1,800 umpires for each phase.

76. Gabel, *GHQ Maneuvers*, 17–19.

77. "Headquarters Armored Force Unit Mobilization Training Program, Sections III and IV (14th to 26th weeks) for Newly Organized Armored Units and Separate Tank Units," 1 May 1942, Box 16, Stack 190, Entry 65, RG 337,

Armored Board, National Archives and Records Administration (hereafter known as NARA II), College Park, Maryland. NARA is located downtown in Washington D.C. The new building in College Park is known as NARA II.

78. Ibid.

79. John K. Waters, SOOHP, MHI, 128.

80. Memorandum AFB 354.2, "Report of Armored Force Board Observer on Louisiana-Maneuvers," 6 October 1941, Box 14, Stack 190, Entry 65, RG 337, Armored Board, NARA II, 2.

81. Ibid., 2–7.

82. "Maneuver Lessons," *Washington Post*, 25 September 1941.

83. Arthur S. Collins Jr., SOOHP, MHI, 57.

84. "Comments by LT. GEN. L. J. McNair, G.S.C., 2d Phase, GHQ—Directed Maneuvers, Camp Polk, LA., September 24–28, 1941," CARL, 5.

85. Memorandum, "Comments on First Phase—Second Army Vs Third Army Maneuvers," 22 September 1941, section 7, 9, CARL.

86. James Carafano, *GI Ingenuity* (Westport, Conn.: Praeger Security International, 2006), 141–163. Carafano tends to credit only the Allied efforts in France for this innovation. However, also see Memorandum, "Report on Close Air Support of the 1st Armored Division," 15 August 1944, CARL. IV Corps in Italy began CAS improvements before the breakout in France. In both cases, tactical commanders developed bottom-up solutions after a year and a half of frustration. Carafano intimates that interwar thought about CAS helped to inspire the innovations in France.

87. Letter from General McNair to General Fuller, "Discussion Louisiana Maneuvers," September 1941, CARL, 41.

88. Donald Bennett, SOOHP, MHI, 20–21.

89. Memorandum, "Comments on First Phase—Second Army Vs. Third Army Maneuvers," 22 September 1941, section 2, Command and Staff, CARL, 2.

90. Ibid.

91. "Development of Army Operational Technique," Memorandum for the President from Army Chief of Staff, 27 March 1943, Box 164, Folder 1, Franklin Delano Roosevelt Presidential Library and Museum (hereafter known as FDRL) at Hyde Park, New York.

92. Walter Kretchik, *US Army Doctrine: From the American Revolution to the War on Terror* (Lawrence: University Press of Kansas, 2011), 143–150.

93. Ibid., 151.

94. General Board, United States Forces, European Theater, "Organization, Equipment, and Tactical Employment of the Armored Division," Study Number 48, 7 November 1945, Center of Military History, Washington, D.C., chapter 3, appendixes 1, 2, and 3; General Board, "Organization, Equipment, and Tactical Employment of the Infantry Division," Study Number 15, 20 November 1945, section 2 and appendix 2; General Board, "Types of Divisions—Post War Army," 20 November 1945, Study Number 17, chapter 3.

95. Mark Watson, *Chief of Staff: Prewar Plans and Preparations* (Washington,

D.C.: U.S. Government Printing Office, 1950), 245–246. In late 1941, the army retired 195 regulars, 142 reserve officers, and 127 National Guard officers, which accounted for approximately 1 percent of the entire officer corps. Henry G. Gole, *The Road to Rainbow: Army Planning for Global War, 1934–1940*, Association of the U.S. Army Book (Annapolis, Md.: Naval Institute Press, 2003), 132; Donald E. Vandergriff, *Culture Wars*, http://www.d-n-i.net/fcs/culture_wars.htm, accessed 28 March 2009.

96. Thomas Griess interview, Tape 20, Side 2, Box 8, 24–25, Jacob Devers Papers, York Heritage Society, York, Pennsylvania; Coffman, *The Regulars*, 415; Robert R. Palmer, Bell I. Wiley, William R. Keast, *The Procurement and Training of Ground Combat Troops: The Army Ground Forces*, CMH Pub. vol. 2-2 (Washington, D.C.: Center of Military History, U. S. Army, 2003), 99–101; Mark A. Stoler, *George C. Marshall: Soldier-Statesman of the American Century* (Boston: Twayne, 1989), 83–84.

97. Classification of Officers—Regular Army Based on General Efficiency Records, 15 March 1937, War Department, Statistics Branch, 1917–1945, Special Statistical Reports, No. 212, 217, 238, 247, RG 165, NARA II,7–9. The War Department developed these confidential reports every fiscal year. They include a summary for the current fiscal year broken down by branch and rank. Furthermore, the report compares the current fiscal year data with the last ten years to analyze trends. I have selected five reports from 1931 to 1939 to evaluate the efficiency reports.

98. I experienced this hyperinflation with the first officer efficiency report I received as a second lieutenant in 1997. Every single officer in our battalion received the highest rating or top block except one who committed several disciplinary infractions. The report had become meaningless since every officer received the same rating.

99. Classification of Officers—Regular Army Based on General Efficiency Records, 15 March 1937, War Department, Statistics Branch, 1917–1945, Special Statistical Reports, No. 238, RG 165, NARA II, 2.

100. Ibid.

101. Relative Efficiency of Officers, 7 January 1936, War Department General Staff Statistics Branch, 1917–1945, Special Statistical Reports, No. 237, RG 165, NARA II, 1–5. In fact, General McNair lent the most weight to these relative rankings when he had to evaluate general officers and colonels for command positions. See Palmer et al., *The Procurement and Training of Ground Combat Troops*, 99.

102. Classification of Officers—Regular Army Based on General Efficiency Records, 2 March 1932, War Department, Statistics Branch, 1917–1945, Special Statistical Reports, No. 217, RG 165, NARA II, 1.

103. Classification of Officers—Regular Army Based on General Efficiency Records, 15 March 1937, War Department, Statistics Branch, 1917–1945, Special Statistical Reports, No. 212, 217, 238, 247, RG 165, NARA II

104. Classification of Officers—Regular Army Based on General Efficiency

Records, 15 March 1937, War Department, Statistics Branch, 1917–1945, Special Statistical Reports, No. 212, 217, 238, 247, RG 165, NARA II, 7–9.

Chapter 3: Wading in with TORCH

1. Stoler, *George C. Marshall*, 96–102; Arthur Layton Funk, *The Politics of TORCH: The Allied Landings and the Algiers Putsch, 1942* (Lawrence: University Press of Kansas, 1974), 65–88.

2. Colin John Bruce, *Invaders: British and American Experience of Seaborne Landings, 1939–1945* (Annapolis, Md.: Naval Institute Press, 1999), 45–49.

3. Ibid., 48–49. George F. Howe, *The Battle History of the 1st Armored Division, "Old Ironsides"* (Washington, D.C.: Combat Forces Press, 1954), 60–91.

4. Gerhard L. Weinberg, "Some Thoughts on World War II," *Journal of Military History* 56 (October 1992): 662–663. Weinberg identifies the French colonial army's propensity to fight Allied forces instead of Axis forces as a major question of World War II that scholars should answer.

5. Carlo D'Este, *Major Battles and Campaigns: World War II in the Mediterranean, 1942–1945* (Chapel Hill, N.C.: Algonquin Books, 1990), 1–22.

6. Howe, *1st Armored*, 89–90.

7. Ralph Francis Bennett, *Ultra and Mediterranean Strategy, 1941–1945* (London: Hamish Hamilton Ltd., 1989), 186.

8. Foreign Military Studies, MS#D066, Generalmajor Christian Eckhard, "Study of the Situation in the High Command of the Wehrmacht Shortly Before, During, and After the Allied Landing in French North Africa, 1942," 3 March 1947, Headquarters, European Command, Office of the Historian Chief, MHI, 7–13.

9. In actuality, the 164th Infantry Regiment from the Americal Division executed the U.S. Army's first combat missions against the Axis on Guadalcanal when they joined the plucky 1st Marine Division. By January 1943, the rest of the division plus the 25th Infantry Division joined them and the fresh 2nd Marine Division, which relieved the 1st MARDIV. I have neglected to include these army units in this study for a few reasons. In order to keep a reasonable scope of study, I decided to remain in one theater. The addition of the marines complicates the issue as well. I believe that I would have to examine marine units to analyze exchange of tactics and techniques between services. Please see *Guadalcanal*, U.S. Army Campaigns of World War II, Center of Military History Publication 72-8, for a brief overview, and Richard Frank, *Guadalcanal* (New York: Penguin Books, 1990), for the most comprehensive campaign account.

10. *Weather Effect on Army Operations: Weather in World War II*, vol. 1, 15 September 1956, CARL, III-4. The Task Force had only 136 remaining serviceable craft out of 378.

11. Report of Operations in North Africa, Memorandum for the Chief of

Staff, Ground Observer Report, 12 December 1942, CARL, 18; Lessons from Operations TORCH, Memorandum from Second Armored Division to Commanding General, Western Task Force, 26 December 1942, CARL, 10.

12. Lessons from Operations TORCH, Memorandum from Second Armored Division to Commanding General, Western Task Force, 26 December 1942, CARL, 3.

13. Ibid., 12; George F. Howe, *Northwest Africa: Seizing the Initiative in the West*, United States Army in World War II: Mediterranean Theater of Operations (Washington, D.C.: Office of the Chief of Military History, Dept. of the Army, 1957), 153–154.

14. Ground Observer Report on the Landing Operations at Port Lyautey, French Morocco, 8–15 November 1942, 30 November 1942, CARL, 3–5; Lessons from Operations TORCH, 19 January 1943, Staff Memorandum #7, CARL, 1–3, 5, 15, 30, 43, 47, 53, 57, 64.

15. Howe, *Northwest Africa*, 122.

16. Report of Operations in North Africa, Memorandum for the Chief of Staff, Ground Observer Report, 12 December 1942, CARL, 3–13.

17. Rick Atkinson, *An Army at Dawn: The War in North Africa, 1942–1943*, 1st ed. (New York: Henry Holt, 2002), 141–158.

18. Reports on Combat Experience and Battle Lessons for Training Purposes, 21 June 1943, Memorandum from Headquarters, Ninth Infantry Division to Commanding General, AFH, Box 7327, RG 407, NARA II, 1–9.

19. Silver Star Citation, Stokes, William M., Lt. Col., Commander of 2nd Battalion, 67th Armored Regiment, 28 November 1942, General Order Number 3, Headquarters Blackstone, William Stokes Papers, George C. Marshall Foundation, Lexington, Virginia. Stokes led his battalion's advance guard ashore to defeat a local French armor counterattack and then advanced inland to defeat another on 9 November; Ground Observer Report on the Landing Operations at Port Lyautey, French Morocco, 8–15 November 1942, 30 November 1942, CARL, 10–11.

20. Report of Operations in North Africa, Memorandum for the Chief of Staff, Ground Observer Report, 12 December 1942, CARL, 22.

21. Report on Observations Made as Observer with Task Force Brushwood, Memorandum to the Commanding General, Army Ground Forces, n.d., Report of Observers: Mediterranean Theater of Operations, MHI, 10.

22. See James Carafano, *GI Ingenuity* (Westport, Conn.: Praeger Security International, 2006), 123–126, for a classic example: the Rhino tank during the Normandy campaign.

23. Report of Operations in North Africa, Memorandum for the Chief of Staff, Ground Observer Report, 12 December 1942, CARL, Enclosure B.

24. William P. Yarborough, *Bail Out over North Africa: America's First Combat Parachute Missions, 1942* (Williamstown, N.J.: Phillips Publications, 1979), 17–20.

25. Edson Raff, *We Jumped to Fight* (New York: Eagle Books, 1944), 21–22.

26. William P. Yarborough, SOOHP, MHI, 8–9.

27. Henry C. Tipton, "The Operations of the 509th Parachute Infantry Battalion (Separate), in the Invasion of North Africa, 8–11 November 1942 (North Africa Campaign)," World War II Monographs Collection, Donovan Research Library (hereafter known as DRL), Ft. Benning, Georgia, 9.

28. Yarborough, SOOHP, MHI, 10–11.

29. Raff, *We Jumped to Fight*, 22–25; Observers' Reports, Airborne Operations in the African Theater to Date, 15 February 1943, Headquarters Army Ground Forces, CARL, 2.

30. William C. Bentley, A Report Discussing Paratroop Operations in the North African Mission, 27 March 1943, CARL, 1–3.

31. Tipton, "The Operations of the 509th PIB," DRL, 9.

32. See Atkinson, *An Army at Dawn*, 87–91, for a brief description of this failed operation. Also, Howe, *Northwest Africa*, 212–213.

33. Lloyd Wilson, "The Operations of the 509th Parachute Battalion in North Africa, 8 November 1942–February 1943," DRL, 11.

34. Raff, *We Jumped to Fight*, 40–46.

35. "U.S. Rangers" by William O. Darby, lecture presented at the Army and Navy Staff College, Washington, D.C., 27 October 1944, CARL, 16–18. This lecture was presented in conjunction with Colonel D. W. Dammer as well as naval officers involved with landings and combat in both North Africa and Sicily. Darby led the discussion, but the other officers contributed significantly.

36. Ibid., 14–15, 17–18.

37. Jerome J. Haggerty, "A History of the Ranger Battalions in World War II" (Ph.D. diss., Fordham University, 1982), 85–87.

38. Combat commands utilized in the U.S. Army throughout the North Africa Campaign consisted of a core regiment, either armored, armored infantry, or light infantry. Under modern-day terminology, one would classify them as regimental combat teams. They typically included supporting artillery, tank destroyer, reconnaissance units, and logistic units. After North Africa, armored divisions were reorganized to have three combat commands, which included one tank battalion, one infantry battalion, and one artillery battalion.

39. Operation VILLAIN was the first American airborne operation of the war.

40. Howe, *Northwest Africa*, 195–211.

41. I use the term LST here to describe the three maracaibo oilers used in this operation. They were the predecessors to the actual LST. However, LST is a very familiar term, and it suggests the purpose of these ships.

42. Lt. Col. Bogardus S. Cairns, "Employment of Armor in the Invasion of Oran," *Military Review* 28 (September 1948): 55–56. The medium M3 Grants had to be unloaded with cranes at loading docks.

43. Richard M. Leighton, Robert W. Coakley, and Center of Military History, *Global Logistics and Strategy, 1940–1943*, CMH Pub. vol. 1-5 (Washington, D.C.: Center of Military History, U.S. Army, 1995). 445–449.

44. R. P. Hunnicutt, *A History of the American Light Tank* (Novato, Calif.: Presidio, 1992), 5–11. Two 25-gallon jettison gas tanks could be added to increase

the range of the Stuart. Crews had to separate these additional tanks prior to combat, and they were practical only for long movements.

45. Howe, *1st Armored*, 24–38.

46. General John K. Waters, SOOHP, MHI, 165–169.

47. Ibid., 169; Orr Kelly, *Meeting the Fox: The Allied Invasion of Africa, from Operation Torch to Kasserine Pass to Victory in Tunisia* (New York: J. Wiley, 2002), 59–60.

48. Kelly, *Meeting the Fox*, 60. In an oral interview with Orr Kelly, Tuck could not remember the name of this French captain, but he remembered running into him again a couple of times in Tunisia. Based on Tuck's assessment, I have assumed that the French officer in question was a competent commander.

49. *FM 17-10, Armored Force Field Manual*, MHI, 92.

50. "The Center Task Force (Oran) in the North African Landing Operation," n.d., William S. Biddle Papers, Box 13, MHI.

51. 1st Armored Regiment, "In the Field," 11 November 1942, 2; John K. Waters, SOOHP, MHI.

52. John Waters married General Patton's daughter, Beatrice. He was captured during the German seizure of Faid Pass and was a POW for the rest of the war. Besides his service in North Africa, he is probably best known as the inspiration for the controversial Hammelburg raid that ended in disaster. Patton sent a mission 40 miles behind enemy lines to try to rescue Waters from Oflag XIIIB. On 26 March 1945, the meager force of 16 tanks, 27 half-tracks, three 105 SP guns, and 294 men under the command of a Captain Abraham Baum reached the camp, never rescued Waters, but helped in getting him wounded by a German paratrooper. While attempting to reach U.S. lines, Task Force Baum was annihilated by three German divisions. The decision to send this force is probably the blackest mark on Patton's record since it was solely to rescue his son-in-law. See Carlo D'Este, *Patton: Genius for War* (New York: Harper Collins, 1995), 714–719, for a brief account of this action.

53. Cairns, "Employment of Armor in the Invasion of Oran," 55–56; Hamilton H. Howze, *A Cavalryman's Story: Memoirs of a Twentieth-Century Army General* (Washington, D.C.: Smithsonian Institution Press, 1996), 47–55; Paul McDonald Robinett, *Armor Command: The Personal Story of a Commander of the 13th Armored Regiment, of the CCB, 1st Armored Division, and of the Armored School during World War II* (Washington, D.C.: McGregor & Werner, 1958), 44–52.

54. Waters, SOOHP, 175–176.

55. Ibid., 170–171.

56. Atkinson, *An Army at Dawn*, 129.

57. James Scott Wheeler, *The Big Red One: America's Legendary 1st Infantry Division from World War I to Desert Storm*, Modern War Studies (Lawrence: University Press of Kansas, 2007), 143–152; Howe, *Northwest Africa*, 215–220.

58. "Lessons of Operation Torch," Staff Memorandum #7, 19 January 1943, Allied Force Headquarters, CARL, 7.

59. Stephen V. Ralph, "The Operations of the 2nd Battalion, 16th Infantry in the Invasion of North Africa, 8–11 November, 1942," DRL, 8–10.

60. TORCH Operation, Operations Summary, 1st Infantry Division, 24 November 1942, World War II Operational Reports, CARL, 2–3.

61. 7th Field Artillery After Action Report, Box 5883, RG 407, NARA II; Wheeler, *The Big Red One*, 143–152; Terry Allen, "Situation and Operations Report of the First Infantry Division, 8 August 1942 to 7 August 1943," CARL, 1–3.

62. Battle of Oran, Combat Team 18, "Action against the Enemy in North African Operations," 19th November 1942, Box 5252, RG 407, NARA II, 1–4.

63. Steven Clay, *Blood and Sacrifice: The History of the 16th Infantry Regiment from the Civil War through the Gulf War* (Wheaton, Ill.: Cantigny First Division Foundation, 2001), 156.

64. Ibid.

65. Field Order #2, 8 November 1942, Field Order #3, 9 November 1942, Field Order #4, 10 November 1942, Headquarters, 1st Infantry Division, World War II Operational Reports, CARL.

66. Appendix Number 4 to Annex Number 2 to Field Order #1, Headquarters, 1st Infantry Division, 10 October 1942, World War II Operational Documents, CARL.

67. G-2 Report, #6740, Germany (Combat), German General Staff School, submitted by Albert Wedemeyer, 3 August 1938, MHI, 48–138.

68. Charles E. Heller and William A. Stofft, *America's First Battles* (Lawrence: University Press of Kansas, 1986), 261–263.

69. "Training: Armored Force Replacement Training Center," 1–9, n.d., Orlando Ward Papers, MHI.

70. Ground Observers Report, first comment by SGT James H. Bowser, Tank Commander, Company H, 1st Armored Regiment, 1st Armored Division (age 23, six years—regular army), second comment by LTC Louis Hightower, 3, 8, United States Army Ground Forces Observer Board, Report of Observers: Mediterranean Theater of Operations, vol. 2, 29 March to 27 November 1943, MHI.

71. Waters, SOOHP, 150.

72. Ibid., 131; Christopher Gabel, *The U.S. Army GHQ Maneuvers of 1941* (New York: Diane Pub. Co., 1992), 191.

73. Howze, *A Cavalryman's Story*, 76.

74. Robinett, *Combat Command*, 16.

75. Waters, SOOHP, 142.

76. Howze, SOOHP, MHI, 88–92; Howze, *A Cavalryman's Story*, 39–42. While Howze is particularly negative on the training of the 1st Armored, his comments reflect a G3 who felt guilty about not preparing the division better. He was thrust into the position and had no real staff experience prior to being assigned as the G3. Furthermore, he totally discounts the five months the

division had to prepare in Ireland. That is a significant amount of time for a unit to possess prior to combat. Also, Howze made these comments after a very successful career in the U.S. Army and probably viewed his time in the 1st Armored as amateurish compared to the advances made in the army, especially in terms of training facilities and officer corps professionalization.

77. Robinett, *Combat Command*, 17, 29; Howze, SOOHP, 85–86; Waters, SOOHP, 143–144.

78. Howe, *1st Armored*, 16.

79. Peter C. Hains III, SOOHP, MHI, 24.

80. Ibid.

81. Robinett, *Combat Command*, 197–199; Howe, *1st Armored*, 199–201.

82. *FM 17-33*, 89, MHI.

83. Atkinson, *An Army at Dawn*, 49–68; Howe, *1st Armored*, 60–93.

84. RESERVIST was similar in nature to TERMINAL except it was designed to take the Oran port facilities. Lieutenant Colonel Marshall led the 3rd Battalion, 6th Armored Infantry, in the assault, which failed completely, suffering almost 90 percent casualties including Marshall. Marshall was last seen leading a small detachment of men hurling grenades at a French destroyer crew.

85. "U.S. Rangers" by William O. Darby.

86. Atkinson, *An Army at Dawn*, 53–54.

87. Ibid., 69–72, 96.

88. Leslie W. Bailey, "The Operations of the 3rd Battalion, 135th Infantry at Algiers, North Africa, 7 November–10 November, 1942," DRL, 6–7.

89. After Action Report, 15 October to 10 November 1942, written by Lt. Col. E. T. Swenson, n.d., 1–2, Box 9567, RG 407, NARA.

90. Bailey, "The Operations of the 3rd Battalion, 135th Infantry," DRL, 12–14.

91. Ibid., 13. The battalion had briefed four contingency plans. Plan A was the successful assault of both destroyers. Plan B was the successful landing of just HMS *Broke*, and Plan C was the landing of only HMS *Malcolm*. Plan D was the evacuation plan for both forces if the destroyers had to leave the harbor due to enemy fire.

92. After Action Report, 15 October to 10 November 1942, written by Lt. Col. E. T. Swenson, n.d., Box 9567, RG 407, NARA II, 6–7.

93. Phase III 135th Infantry Regiment, "The Landing at Algeria," John M. Briet Papers, Iowa Gold Star Military Museum, Johnston, Iowa, 4–5.

94. Bailey, "The Operations of the 3rd Battalion, 135th Infantry," DRL, 27. Leslie Bailey was a platoon leader during the TERMINAL operation. He emphasized that all the leadership had a complete grasp of the mission based on the orders and rehearsals conducted aboard the ships. Most of his monograph is based on his personal experiences. The rapid assault and seizure of the objectives confirm subordinate leader comprehension.

95. After Action Report, 15 October to 10 November 1942, written by Lt. Col. E. T. Swenson, n.d., Box 9567, RG 407, NARA II, 7–8.

96. Report on Impressions Gained from the Assault Phase of the Operations in North Africa by Major General J. C. Haydon, n.d., CARL, 12–13.

97. Ibid., 4–7, 12–13, 16.

Chapter 4: The Race for Tunisia

1. Weather Effect on Army Operations, Weather in World War II, vol. 1, CARL, III-2.

2. Observer Report, 13 March 1943, Army Ground Forces, CARL, 8; George F. Howe, *Northwest Africa: Seizing the Initiative in the West*, United States Army in World War II: Mediterranean Theater of Operations (Washington, D.C.: Office of the Chief of Military History, Dept. of the Army, 1957), 291–292.

3. Michael King, *William Orlando Darby: A Military Biography* (Hamden, Conn.: Archon Book, 1981), 69.

4. William P. Yarborough, *Bail Out over North Africa: America's First Combat Parachute Missions, 1942* (Williamstown, N.J.: Phillips Publications, 1979), 137.

5. Robert R. Palmer, Bell I. Wiley, and William R. Keast, *The Procurement and Training of Ground Combat Troops: The Army Ground Forces*, CMH Pub. vol. 2-2 (Washington, D.C.: Center of Military History, U. S. Army, 2003), 6–8.

6. Ibid., 16.

7. Some officers have concluded that intellectual ability is critical to effective and aggressive leaders on the battlefield. Furthermore, they conclude that officers traditionally possess the ability to outperform their subordinates in combat based on their higher levels of acumen. See Thomas A. Horner, "Killers, Fillers, and Fodder," *Parameters* 12, no. 3 (September 1982): 27–34, for an example of this.

8. Ibid., 4–5.

9. Ibid., 4–32.

10. Peter R. Mansoor, *The GI Offensive in Europe: The Triumph of American Infantry Divisions, 1941–1945* (Lawrence: University Press of Kansas, 1999), 249–267.

11. Observer Report, 13 March 1943, Army Ground Forces, CARL, 1.

12. William R. Keast, *Provision of Enlisted Replacements*, Study No. 7, Historical Section, Army Ground Forces, 1946, U.S. Army Center of Military History, Washington, D.C., 7; Observer Report, 13 March 1943, Army Ground Forces, CARL, 1.

13. Observer Report, 13 March 1943, Army Ground Forces, CARL, 8–11.

14. William R. Keast, *The Procurement and Branch Distribution of Officers*, Study No. 4, Historical Section, Army Ground Forces, 1946, USACMH, 4.

15. See Cole Kingseed, *Old Glory Stories: American Combat Leadership in World War II* (Annapolis, Md.: Naval Institute Press, 2006), 173–191, 246–264, for brief combat records or the memoirs that he edited for more detailed discussion. Kingseed developed close relationships with both Winters and

Dawson and edited their memoirs. See Dick Winters, *Beyond Band of Brothers: The War Memoirs of Dick Winters* (New York: Berkley Caliber, 2006), and Cole Kingseed and Joe Dawson, *From Omaha Beach to Dawson's Ridge: The Combat Journal of Captain Joe Dawson* (Annapolis, Md.: Naval Institute Press, 2005).

16. See Charles B. MacDonald, *Company Commander* (New York: Bantam Books, 1947), for MacDonald's honest self-assessment and one of the best company memoirs of the war. Murphy is still a legend even in the modern U.S. Army. An entire NCO competition is named in his honor.

17. Horner, "Killers, Fillers, and Fodders," 29–32.

18. Observer Report, 13 March 1943, Army Ground Forces, CARL, 4.

19. Ralph Francis Bennett, *Ultra and Mediterranean Strategy* (London: Hamish Hamilton, 1989), 402–403.

20. See Jean E. Engler, SOOHP, MHI, Tape 1: 43–48, for one example, or Memorandum AFB 354.2, "Report of Armored Force Board Observer on Louisiana-Maneuvers," 6 October 1941, Box 14, Stack 190, Entry 65, RG 337, Armored Board, NARA II.

21. Lieutenant Colonel B. A. Dickson, Observer Report, 30 December 1942, Army Ground Forces, CARL, 3.

22. Observer Report, 13 March 1943, Army Ground Forces, CARL, 3.

23. Daniel R. Mortensen, *A Pattern for Joint Operations: World War II Close Air Support North Africa* (Washington, D.C.: Office of Air Force History and U.S. Army Center of Military History, 1987), 59–67.

24. E. A. Russell, Observer Report, 19 February 1943, Army Ground Forces, CARL, 2.

25. Ibid.

26. *FM 100-20, Command and Employment of Air Power* (Washington, D.C.: U.S. Government Printing Office, 1943), 4. The Germans thought the same thing. As they slowly lost air superiority in North Africa, the Wehrmacht lost the ability to maneuver and entered into a state of *Stellungskrieg*. Essentially, "Stellungskrieg" means position warfare, and it forced the Germans to adopt a defensive posture with the loss of the "third dimension: airpower." Robert Michael Citino, *Death of the Wehrmacht: The German Campaigns of 1942*, Modern War Studies (Lawrence: University Press of Kansas, 2007), 213–222.

27. The majority of German tanks in theater were the Mark IIIs and IVs. While no definitive figure exists to determine the exact number of Tigers, there were probably no more than 25–30 of these tanks. This estimation is based on Ultra information, personal memoirs, and German logbooks. See Bennett, *Ultra and Mediterranean Strategy*, 157, 194–196.

28. *Tunisia*, U.S. Army Campaigns of World War II, U.S. Army Center of Military History, Publication 72-12, n.d., 4, 11.

29. Freeland A. Daubin, "The Battle of 'Happy Valley,'" 24 April 1948, MHI, 9.

30. *First Battalion, First Armored Regiment's Participation in the Beginning of the North African Campaign*, 31 December 1942, Waters Papers, MHI.

31. Daubin, "The Battle of 'Happy Valley,'" 17.

32. David Rolf, *The Bloody Road to Tunis: Destruction of the Axis Forces in North Africa* (London: Greenhill Books, 2001), 50–51.

33. Daubin, "The Battle of 'Happy Valley,'" 23.

34. Waters, SOOHP, 152. *FM 17-33, Armored Force Field Manual: The Armored Battalion, Light and Medium* (Washington, D.C.: U.S. Government Printing Office, 1942), 66, 97, 92.

35. Daubin, "The Battle of 'Happy Valley,'" 25; Waters, SOOHP, 153; Kelly, *Meeting the Fox*, 96.

36. Rolf, *The Bloody Road to Tunis*, 51; Walter Nehring, "The First Phase of the Battle of Tunisia," Foreign Military Studies (known hereafter as FMS), D-147, U.S. Army European Command, 1947, 15, MHI.

37. Daubin, "The Battle of 'Happy Valley,'" 28–30.

38. "Brew-up" is a slang term that tankers use to describe a tank when it catches fire and gives off secondary explosions from its onboard ammunition and fuel. Americans adopted this term from the British.

39. Daubin, "The Battle of 'Happy Valley,'" 24–31; Rolf, *The Bloody Road to Tunis*, 5; Kelly, *Meeting the Fox*, 97–99; Howze, SOOHP, 67–68; Waters, SOOHP, 184–185; *First Battalion, First Armored Regiment's Participation in the Beginning of the North African Campaign*, 31 December 1942, Waters Papers, MHI.

40. Report on Combat Experiences and Battle Lessons for Training Purposes, 13 June 1943, Headquarters 1st Armored Division, Folder 2, Pritchard Collection, George S. Patton Museum, Fort Knox, Kentucky (hereafter referred to as GPM), 1–2.

41. Report on Combat Experiences and Battle Lessons for Training Purposes, 6th Armored Infantry Regiment, 10 June 1943, Folder 2, Pritchard Collection, GPM, 4; Report on Combat Experiences and Battle Lessons for Training Purposes, 81st Armored Reconnaissance Battalion, 10 June 1943, Folder 2, Pritchard Collection, GPM 1. These two documents are representative of most lessons-learned documents that stressed mentorship and junior leader development. All commanders recognized their importance and stressed them up and down the chain of command.

42. Observer Report, 5 March 1943, Army Ground Forces, CARL, 18; Observer Report, Lt. Col. W. H. Schaefer and Maj. Franklin Gardner, 10 February 1943, Army Ground Forces, CARL, 6.

43. Gardiner Papers, USMA, 89–90.

44. Howe, *Northwest Africa*, 326.

45. Diary, 1st Battalion, 6th Armored Infantry, North African Campaign, 1942–1943, Paul Robinett Papers, GMF, 4.

46. Observer Report, Lt. Col. W. B. Kern, 19 February 1943, Army Ground Forces, CARL, 1; Observer Report, Lt. Col. E. A. Russell, January 1943, Army Ground Forces, CARL, 2.

47. Diary, 1st Battalion, 6th Armored Infantry, North African Campaign, 1942–1943, Paul Robinett Papers, GMF, 4–7; George F. Howe, *The Battle*

History of the 1st Armored Division, "Old Ironsides" (Washington, D.C.: Combat Forces Press, 1954), 82–85.

48. Howe, *Old Ironsides*, 91–92.

49. Observer Report, 5 March 1943, Army Ground Forces, CARL, 16–17.

50. Obituary, John Robert McGinness, *Assembly* 33, no. 2 (September 1974): 122–123.

51. Observer Report, 5 March 1943, Army Ground Forces, CARL, 16–17; TORCH Summaries, 1900 December 13 to 0700 December 14, Box 108, Map Room Files, FDR.

52. Diary, 1st Battalion, 6th Armored Infantry, North African Campaign, 1942–1943, Paul Robinett Papers, GMF, 4–8.

53. Reports on Combat Experience and Battle Lessons for Training Purposes, 1st Armored Regiment, 9 June 1943, Folder 1, Pritchard Collection, GPM, 1. This is just one example of many. Most maneuver commanders recognized the absolute necessity to combine infantry and tank units at least at the battalion, if not company, level.

54. Operations of the 81st Armored Reconnaissance Battalion in Tunisia, Cavalry Reconnaissance, Number One, Cavalry School, 1943, CARL, 1, 8.

55. The term "counter-battery fire" refers to a unit using its artillery to fire on the enemy's artillery. This tactic requires great precision and is not easy.

56. Report on Combat Experiences and Battle Lessons for Training Purposes, 91st Armored Field Artillery Battalion, 10 June 1943, Folder 2, Pritchard Collection, GPM, 1.

57. William R. Keast, *Provision of Enlisted Replacements*, Study No. 7, Historical Section, Army Ground Forces, 1946, USACMH, 16; Mansoor, *The GI Offensive*, 250–251.

58. James Carafano, *GI Ingenuity* (Westport, Conn.: Praeger Security International, 2006), 198–201.

59. Howe, *Old Ironsides*, 93.

60. Robinett, *Armor Command*, 155.

61. Observer Report, Lt. Col. W. H. Schaefer and Maj. Franklin Gardner, 10 February 1943, Army Ground Forces, CARL, 2.

62. Report of Longstop Hill Engagement, Tunisia, 20 March 1943, Box 5252, RG 407, NARA II, 1.

63. Foreign Military Studies, MS#D173, Rudolf Lang, "Battles of Kampfgruppe Lang in Tunisia December 1942 to 15 April 1943, Part I," n.d., Headquarters, European Command, Office of the Historian Chief, MHI, 1–7.

64. Robert W. Porter, SOOHP, MHI, 228–229.

65. Report of Longstop Hill Engagement, Tunisia, 20 March 1943, Box 5252, RG 407, NARA II, 2.

66. Rick Atkinson, *An Army at Dawn: The War in North Africa, 1942–1943*, 1st ed. (New York: Henry Holt, 2002), 253–255.

67. Report of Longstop Hill Engagement, Tunisia, 20 March 1943, Box 5252, RG 407, NARA II, 2.

68. James Scott Wheeler, *The Big Red One: America's Legendary 1st Infantry Division from World War I to Desert Storm*, Modern War Studies (Lawrence: University Press of Kansas, 2007), 162.

69. Report of Longstop Hill Engagement, Tunisia, 20 March 1943, Box 5252, RG 407, NARA II, 3.

70. Rudolf Lang, "Battles of Kampfgruppe Lang in Tunisia December 1942 to 15 April 1943, Part I," 8–9.

71. Mortensen, *A Pattern for Joint Operations*, 59.

72. Hamilton H. Howze, *A Cavalryman's Story: Memoirs of a Twentieth-Century Army General* (Washington, D.C.: Smithsonian Institution Press, 1996), 56–57; *Tunisia*, The U.S. Army Campaigns of World War II, USACMH, Publication 72-12, n.d., 10–11.

73. Observer Report, Lt. Col. W. H. Schaefer and Maj. Franklin Gardner, 10 February 1943, Army Ground Forces, CARL, 5–6; Unit History of the Ninth Infantry Division for the Year 1943, 2 May 1944, Box 7325, RG 407, NARA II, 1–3.

74. Department of the Scientific Adviser to the Army Council, Military Operational Research Report, Report #33, "Tank Battle Analysis," November 1946, MHI.

75. "Reports on Combat Experience and Battle Lessons for Training Purposes," Memorandum to Commanding General, 1st U.S. Armored Division from Lt. Col. E. A. Russell, Executive Officer for CCB, 1st Armored Division, 10 June 1943, Box 1, Folder 2, Pritchard Collection, Patton Museum, 1. This is just one example, but all the lessons-learned documents are riddled with discussions on how to best coordinate infantry and tanks in the attack.

76. Daubin, "The Battle of 'Happy Valley,'" 17–19.

77. *Survey of Allied Tank Casualties in World War II*, MHI, 14.

78. Ibid., 35.

79. Ibid., 92.

80. The ME-109 and Junker 87 (Stuka) were the most ubiquitous aircraft in North Africa. Their machine armaments were usually quite small (8mm) compared to the .50 calibers on some Allied aircraft. Unless a direct hit from a dive-bombing Stuka destroyed a tank, it is unlikely that these smaller calibers would penetrate the armor of any U.S. tanks.

81. Citino, *Death of the Wehrmacht*, 212–222.

82. Erwin Rommel, *Rommel Papers*, ed. B. H. Liddell-Hart (New York: Harcourt, Brace, 1953), 283–285.

83. Maj. Gen. E. N. Harmon, Notes on Combat Experience during the Tunisian and African Campaigns, 4, *Kasserine Pass Battles*, vol. 2, part 3, USACMH; Harmon, *Combat Commander*, 146; George Forty, *United States Tanks of World War II in Action* (New York: Blandford Press, 1983), 134.

84. Belton Cooper, *Death Traps* (New York: Ballantine Books, 1998), 24–30.

85. Forty, *United States Tanks*, 124–127.

86. Army Ground Force Equipment Review Board, Part I, Records of the

War Department General and Special Staffs—Equipment Board, 20 June 1945, Box 200, RG 165, NARA II, 81–87.

87. Detailed Comment by the Army Ground Forces on Final Report of Ground Force Equipment Review Board, n.d., Records of the War Department General and Special Staffs—Equipment Board, Box 200, RG 165, NARA II.

88. Army Ground Force Equipment Review Board, Part I, 13–16.

89. On four separate matters, general officers specifically asked for Hightower's opinion. On no other issue did board leaders ask for another subordinate's opinion.

90. War Department Equipment Board: Armored Hearings, vol. 7, Armored and Tank Destroyer, 4–6 December 1945, Box 200, RG 165, NARA II, 4–9.

91. Ibid., 1–15.

92. Ibid., 1, 17–19.

93. Ibid., 39.

Chapter 5: A Long February in the Tunisian Passes

1. George F. Howe, *Northwest Africa: Seizing the Initiative in the West*, United States Army in World War II: Mediterranean Theater of Operations (Washington, D.C.: Office of the Chief of Military History, Department of the Army, 1957), 388–393; Carlo D'Este, *World War II in the Mediterranean, 1942–1945* (Chapel Hill, N.C.: Algonquin Books, 1990), 12–16.

2. Martin Blumenson, *Kasserine Pass* (Boston: Houghton Mifflin, 1967), 131–133.

3. Bennett, *Ultra and Mediterranean Strategy*, 200.

4. Howe, *Northwest Africa*, 401.

5. "Djebel" means mountain in Arabic and is abbreviated throughout U.S. documents as DJ. Allied commanders used Djebel since all of their maps used the term or its abbreviation.

6. "Defense of Faid Position," from Fredendall to Ward, 11 February 1943, Orlando Ward Papers, MHI.

7. Waters, SOOHP, MHI, 190.

8. Lieutenant Colonel Waters called him deadwood after the war. McQuillin was relieved shortly after the Kasserine Pass battles due to his incompetence.

9. Waters, SOOHP, 191. McQuillin reassured Waters that no attack would come through Faid Pass. He told Waters that everyone was predicting a German attack in the north.

10. Waters knew that the German infantry would search the hills. He ordered his driver and an assistant captain to break down and hide the radios from the half-track. His driver got shot by jumpy U.S. infantryman and bled to death. Waters told his captain to go hide in one wadi while Waters hid in a

different one. At about 1600, Waters heard someone approaching and assumed that it was his captain coming up the wadi. He stepped into the open and saw a group of Arabs leading a German infantry patrol up the trail. The patrol spotted Waters and shot a few rounds at him even though he was trying to surrender. Luckily for Waters, the rounds missed him. The Arabs had led the German patrol right to Waters's position, and he had no opportunity to escape. From Waters, SOOHP, 210–212.

11. Thomas Drake, "Factual Account of Operations 168th Infantry, 34th Division, from 24 December 1942 to 17 February 1943," 2 April 1945, Iowa Gold Star Military Museum, 2–5.

12. Rick Atkinson, *An Army at Dawn: The War in North Africa, 1942–1943*, 1st ed. (New York: Henry Holt, 2002), 312–317.

13. "History of the 168th Infantry for Period November 12, 1942 to March 15, 1943," Box 9576, RG 407, NARA II, 4–9.

14. Jack Lake, "The Operations of the 168th Infantry, 2d Battalion (34th Division) Faid Pass, 12–21 February 1943 (Tunisian Campaign)," DRL, 19–20.

15. "History of the 168th Infantry for Period November 12, 1942 to March 15, 1943," 17–19.

16. John Van Vliet Jr., "Brief of Statement of Lt. Colonel John H. van Vliet, Jr.," 2 April 1945, Iowa Gold Star Military Museum, 1–2.

17. Ibid., 23–26; Atkinson, *An Army at Dawn*, 355–356.

18. Wadis typically occur in arid regions that sometimes experience flash floods or heavy rainfalls. The deluge of water carves irregular ditches out of the dry soil. The presence of these washes or wadis creates countless terrain features that can hide numerous vehicles or soldiers. Also, the steepness and presence of water prevents easy cross-country movement for vehicles if they are not on roads.

19. See the following volumes for more information on this action: Howe, *Northwest Africa*, 411–415; Atkinson, *An Army at Dawn*, 340–342; Blumenson, *Kasserine Pass*, 149–153; George F. Howe, *The Battle History of the 1st Armored Division, "Old Ironsides"* (Washington, D.C.: Combat Forces Press, 1954), 143–165. Kasserine Pass is actually named for the small town of Kasserine about 5 miles east of the pass in a valley floor.

20. Lt. Col. Louis Hightower, Reports on Combat Experience and Battle Lessons for Training Purposes, June 9, 1943, 1–3, NARA II. A lessons-learned document, known today in the military vernacular as an after action report (AAR), serves as a training tool to capture good and bad experiences on the field of battle. These reports filter up the chain of command so the lessons become nested and mutually supporting. These lessons later affect doctrine, planning, equipment development and allocation, and logistical support, just to name a few of their applications.

21. 1st Armored had two permanent combat commands, but often created a Command C or D as the enemy situation dictated.

22. The ideas running through Hightower's mind are the author's

speculation, based on personal combat experiences as an armor officer in the army in Iraq, Bosnia, and Macedonia, and countless training exercises.

23. Hightower, Reports on Combat Experience and Battle Lessons for Training Purposes.

24. According to the U.S. Army *Field Manual 3-0, Operations*, June 2001, "Doctrine is the concise expression of how Army forces contribute to unified action in campaigns, major operations, battles and engagements. While it complements joint doctrine, Army doctrine also describes the Army's approach and contributions to full spectrum operations on land. Army doctrine is authoritative but not prescriptive. Where conflicts between Army and joint doctrine arise, joint doctrine takes precedence." Joint doctrine includes coordination among the different branches of the service. In the case of North Africa, the coordination of the U.S. Navy and U.S. Army Air Forces with U.S. Army ground forces should give one the idea of how operations get more complex as one climbs the ladder of command. While the official definition of doctrine is a sound starting point, I have tried to explain how an officer learns this doctrine and adapts it. The explanation of doctrine stems from my own personal experiences in the U.S. Army, and I claim responsibility for all terms and definitions in the text. However, several works provide more detailed explanations and some of them are: Williamson Murray and Allan R. Millett, *A War To Be Won* (Cambridge, Mass.: Harvard University Press, 2000); William B. Skelton, *An American Profession of Arms: The Army Officer Corps, 1784–1861* (Lawrence: University Press of Kansas, 1992); Allan R. Millett and Peter Maslowski, *For the Common Defense: A Military History of the United States of America* (New York: Free Press, 1994); and for an account of marine doctrine, read Allan R. Millett, *Semper Fidelis: The History of the United States Marine Corps* (New York: Free Press, 1991).

25. Orlando Ward Papers, Box 5, MHI; see Murray and Millett, *A War To Be Won*, appendixes 1–4, for definitions.

26. Howe, *Northwest Africa*, 411–415; Atkinson, *An Army at Dawn*, 340–342; Blumenson, *Kasserine Pass*, 149–153; Howe, *"Old Ironsides,"* 143–165.

27. Again, I am speculating on Hightower's exact words, but I base this on combat experience and detailed accounts of the fight provided by Hightower himself. See "Single U.S. Tank fights 10 Germans," *New York Times*, 21 February 1943.

28. Operations of 3rd Battalion, 1st Armored Regiment from January 1, 1943 to February 21, 1943, 30 July 1943, RG 407, NARA II; the following titles also have summaries of this action: Atkinson, *An Army at Dawn*; Blumenson, *Kasserine Pass*; Howe, *"Old Ironsides"*; Howe, *Northwest Africa*; Charles Messenger, *The Tunisian Campaign* (London: Ian Allan, 1982).

29. Reports on Combat Experience and Battle Lessons for Training Purposes, June 9, 1943, RG 407, NARA II.

30. Ibid.

31. Ibid.

32. *Kasserine Pass Battles, Readings*, vol. 1, part 1 and part 2, USACMH, Washington, D.C. Created for a battlefield staff ride conducted in 1953.

33. *FM 17-33, Armored Force Field Manual, The Armored Battalion, Light and Medium* (Washington, D.C.: U.S. Government Printing Office, 1942), 75.

34. *FM 17-10, Armored Force Field Manual, Tactics and Technique* (Washington, D.C.: U.S. Government Printing Office, 1942), 259.

35. Maj. Gen. E. N. Harmon, "Commanding General's Report on Combat Experience of 1st Armored Division," 13 June 1943, Box 1, Folder 2, James R. Pritchard Collection, Patton Museum, 12.

36. *FM 21-6, List of Publications for Training, Including Training Films and Film Strips* (Washington, D.C.: U.S. Government Printing Office, 1942), 15–40. The U.S. Army had employed the 75mm since World War I with only slight modifications. In fact, West Point cadets used them to practice artillery gun drills at Trophy Point as late as 1941.

37. Cavalry School, *Cavalry Reconnaissance*, nos. 1–7 (Fort Riley, Kans.: Cavalry School, 1943), CARL.

38. Ibid.; Robert Cameron, *To Fight or Not to Fight?* (Fort Leavenworth, Kans.: Combat Studies Institute Press, 2010), 52–56; George F. Hofmann, *Through Mobility We Conquer: The Mechanization of U.S. Cavalry* (Lexington: University Press of Kentucky, 2006), 299–303; "The 81st Armored Reconnaissance Battalion," Mateur, 3 to 11 May 1943, and 81st Reconnaissance Battalion Operations Log, Box 1, John F. Davis Papers, USMA.

39. Letter from Lt. Gen. Patton to Gen. Marshall, 29 March 1943, Map Room Files, Box 167, FDR Library.

40. Operations of 3rd Battalion, 1st Armored Regiment from 1 January 1943 to 21 February 1943, 30 July 1943, RG 407, NARA II.

41. T. J. Camp, ed., *Tankers in Tunisia* (Fort Knox, Ky.: Armored Replacement Training Center, 1943), 25.

42. "The Tank Battle at Sidi Bou Zid," B19, Paul Robinett Papers, MHI. This document can also be found on the website of the Center for Military History, Washington, D.C.

43. "Fire and maneuver tactics" refers to a common tactic understood by soldiers of combat arms units. This term means to use suppressive fire against an enemy position while another friendly unit maneuvers to the left or right to outflank the enemy force.

44. "The Tank Battle at Sidi Bou Zid," B19, Paul Robinett Papers, MHI.

45. Ibid., B25.

46. Ibid., B23.

47. "Single U.S. Tank Fights 10 Germans," *New York Times*, 21 February 1943.

48. Letter to Colonel Hamilton Howze from LTC John Hightower, 1 July 1946, Howze Papers, MHI.

49. "The Tank Battle at Sidi Bou Zid," B19.

50. Operations of 3rd Battalion, 1st Armored Regiment, From 1 January 1943 to 21 February 1943, 30 July 1943, RG 407, NARA II.

51. *Kasserine Pass Battles*, vol. 1, part 2, 1st Armored Division, Report of Operations, 19–25 February 1943, USACMH.

52. LTG James D. Alger Papers, 10 January 1991, USMA, 8.

53. Alger Collection, USMA, 8.

54. Pichon is near Fondouk Pass, and these units were conducting reconnaissance. Paul McDonald Robinett, *Armor Command: The Personal Story of a Commander of the 13th Armored Regiment, of the CCB, 1st Armored Division, and of the Armored School during World War II* (Washington, D.C.: McGregor & Werner, 1958), 157.

55. Ibid., 163.

56. Alger Collection, USMA, 9–10. See the following volumes for detailed accounts of the battle: Howe, *"Old Ironsides"*; Howe, *Northwest Africa*; see *Tankers in Tunisia*, 25, for more details on German camouflage tactics.

57. The term "button-up" refers to a tank crew closing all of its hatches. This posture affords maximum protection for the crew but greatly reduces visibility. Tank commanders don't like to button-up and do it only when survival dictates.

58. Alger Collection, USMA, 9–12; W. G. F. Jackson, *The North African Campaign 1940–43* (London: B. T. Batsford, 1943), 247–260; Atkinson, *An Army at Dawn*, 350–355; Blumenson, *Kasserine Pass.*

59. Atkinson, *An Army at Dawn*, 350–355.

60. Alger Collection, USMA, 9–14.

61. Ibid., 5.

62. *Kasserine Pass Battles*, vol. 1, part 2, 10th Panzer Division, War Diary, 14–22 February 1943, 21st Panzer Division, War Diary, 14–23 February 1943, USACMH.

63. Kurt E. Wolff, description of battle of Sidi Bou Zid in *Das Reich*, 11 April 1943, translated at the Command and General Staff School, Fort Leavenworth, Kansas, from Tactics Department, Armored School, Fort Knox, Kentucky, December 1943, 4.

64. Ibid., 5.

65. Sidi Bou Zid Battlefield Survey, no date, 1, Orlando Ward Papers, MHI; Department of the Scientific Adviser to the Army Council, Military Operational Research Report, Report #33, "Tank Battle Analysis," November 1946, MHI; "Survey of Allied Tank Casualties in World War II," Technical Memorandum, 31 March 1951, MHI.

66. Orlando Ward Papers, Box 8, MHI.

67. Orlando Ward Papers, Box 10, 16 February 1943, MHI.

68. Letter from Capt. Robert Van Zant to Mrs. Ethel Van Zant, 8 March 1943, Box 2, Van Zant Collection, Patton Museum, 1–3.

69. Henry E. Gardiner, "We Fought at Kasserine," *Armored Cavalry Journal*, March–April, 1948, 11–15; Henry Gardiner Papers, USMA, 130–131.

70. Terry Allen, "Situation and Operations Report of the First Infantry Division during the Period of Its Overseas Movement, North African, and Sicilian Campaigns from 8 August 1942 to 7 August 1943," CARL, 3–4.

71. James Scott Wheeler, *The Big Red One: America's Legendary 1st Infantry Division from World War I to Desert Storm*, Modern War Studies (Lawrence: University Press of Kansas, 2007), 177.

72. "Through the Years with the Sixteenth Infantry," Regimental History, Box 5231, RG 407, NARA II, 34–41; Wheeler, *The Big Red One*, 177–180.

73. "Through the Years with the Sixteenth Infantry," 35.

74. Lloyd Fredendall, "Notes on Recent Operations on the Tunisian Front," 10 March 1943, N6534A, CARL, 1–3.

75. Ground Observers Report, 2, United States Army Ground Forces Observer Board, Report of Observers: Mediterranean Theater of Operations, vol. 2, 29 March to 27 November 1943, MHI.

Chapter 6: A Greenless Spring to End Tunisia

1. Letter, Russell F. Akers, Jr. to Lt. Col. C. B. Hansen 12 February 1951. This is part of Hansen's research draft of *A Soldier's Story*, MHI.

2. "Djebel" means mountain in Arabic and is abbreviated throughout U.S. documents as DJ. Allied commanders used Djebel since all of their maps used the term or its abbreviation.

3. Ward Diary, entry for 24 March 1943, Orlando Ward Papers, MHI.

4. George S. Patton, *The Patton Papers*, comp. and ed. Martin Blumenson (Boston: Houghton Mifflin, 1972–1974), 2:215.

5. Lang, "Report on the Fighting of Kampfgruppe Lang," part 2, 5–8, MHI; Kriegstagebuch V, Fifth Panzer Army, March 23–24, 1943, RG 319, NARA II.

6. See Russell Gugeler's unpublished biography of Orlando Ward, 132–144, Orlando Ward Papers, MHI, for a detailed and balanced account of the conditions surrounding Ward's relief.

7. Robert H. Berlin, "United States Army World War II Corps Commanders: A Composite Biography," *Journal of Military History* 53 (April 1989): 147–150.

8. During this time, General George S. Patton became the new commander of U.S. II Corps on 6 March 1943. He replaced Lieutenant General Lloyd R. Fredendall. Eisenhower had considered relieving General Orlando Ward as well, but he decided that Fredendall was the man to go.

9. George F. Howe, *The Battle History of the 1st Armored Division, "Old Ironsides"* (Washington, D.C.: Combat Forces Press, 1954), 199–209.

10. *Report of Operations, 1st Armored Division, Maknassy, Tunisia*, 20 October 1943, MHI, 4–5.

11. Howe, *"Old Ironsides,"* 210–215.

12. Rick Atkinson, *An Army at Dawn: The War in North Africa, 1942–1943*, 1st ed. (New York: Henry Holt, 2002), 459.

13. James Scott Wheeler, *The Big Red One: America's Legendary 1st Infantry*

Division from World War I to Desert Storm, Modern War Studies (Lawrence: University Press of Kansas, 2007), 192–196.

14. Lessons Learned in Combat, 7–8 November 1942 to September 1944, 34th Infantry Division, Box 2, RG 337 NARA II, chapter 3a, b, and c; "Report on Operation Conducted by 9th Infantry Division, Southern Tunisia, 26 March–8 April 1943," 5 October 1943, Box 7227, RG 407, NARA II; "Report on Operation Conducted by 9th Infantry Division, Northern Tunisia, 11 April–8 May 1943," 10 September 1943, N1479, CARL; "Operations of 18th Infantry in Mateur Sector, Tunisia, April 12, 1943, to May 8, 1943," n.d., Box 5253, RG 407, NARA II, 1–2.

15. Steven Clay, *Blood and Sacrifice: The History of the 16th Infantry Regiment from the Civil War through the Gulf War* (Wheaton, Ill.: Cantigny First Division Foundation, 2001), 167–169.

16. William Darby, Report of Action, 1st Ranger Battalion, 9 April 1943, N17439.4, CARL, 1.

17. Report of Operations, Headquarters Ninth Infantry Division, 25 August 1943, Box 7326, RG 407, NARA II, 10.

18. Ibid., 11.

19. Ibid., 1, 2, 5.

20. Howe, *"Old Ironsides,"* 216–222.

21. Howze, *Tank Action*, 1, Howze Papers, MHI. Howze intended this document for family consumption, so one can assume that he was fairly honest and did not write the piece for publication or as an official report. However, other professionals did see it after he wrote it, so he must have gotten some feedback on it. Regardless, the tone of his document is sincere. Howze came from a well-established military family. His father, Robert Lee Howze, was awarded the Medal of Honor for a small action against the Sioux, fought in the Spanish-American War, and served as the superintendent of West Point. Robert Howze died in Columbus, Ohio, during a gallbladder operation.

22. See Hamilton H. Howze, *A Cavalryman's Story: Memoirs of a Twentieth-Century Army General* (Washington, D.C.: Smithsonian Institution Press, 1996), 233–257, for a description of the entire board's findings. Also see the *U.S. Army Tactical Mobility Requirements Board, Final Report, 20 August 1962*, MHI. The report is 296 pages, but contains an executive summary at the beginning. Although the final report came out in 1962, Howze and his team had worked on the project since its beginning in early 1957.

23. Howze, *A Cavalryman's Story*, 47–48.

24. Not an uncommon practice; generals often transferred key staff officers with them so they did not have to train new personnel to their style of command, especially in the midst of combat.

25. Howze, *Tank Action*, 1; Hamilton Howze, SOOHP, MHI, 146.

26. Lt. Col. Charles J. Hoy, "Reconnaissance Lessons from Tunisia," *Cavalry Journal*, November–December 1943, 16–20. Hoy does a superb job of capturing

the lessons learned, and much of it is in line with current U.S. Army doctrine. He had a clear grasp of the tenets of reconnaissance.

27. Lt. Col. Charles J. Hoy, "Mechanics of Battlefield Reconnaissance," *Cavalry Journal*, May–June, 1944, 24–29.

28. When Hoy returned to the United States after his time in North Africa, he wrote three succinct and poignant articles that bear out his comprehension of armored reconnaissance. See footnotes 26 and 27 above for two examples.

29. "Lessons Learned in Tunisian Campaign," 8 June 1943, RG 407, NARA II.

30. Tanks were not organic to the reconnaissance battalion. However, the light tank battalions provided tank companies or platoons to augment the combat power of the reconnaissance battalions. In fact, reconnaissance became the only role suitable for the lightly armored Stuart tanks.

31. Howe, *"Old Ironsides,"* 100–249.

32. "Abteilung" is a very general term that the German army loosely applied to many composite units. "Kampfgruppe" is another term used similarly. The unit in this case was larger than a battalion, but smaller than a brigade. It usually had about eighty tanks and several batteries of antitank guns. The term is translated as a section, department, detachment, unit, or battalion, according to *Vocabulary of German Military Terms and Abbreviations* (London: War Office, 1943), 8.

33. Matthew Darlington Morton, "Men on 'Iron Ponies": The Death and Rebirth of the Modern U.S. Cavalry" (Ph.D. diss., Florida State University, 2004), 295–298; Toppe, *Desert Warfare*, FMS, MHI, 40.

34. Howze, *Tank Action*, 2–3; Henry E. Gardiner Papers, USMA, 52–60.

35. At this time in Tunisia, most of the tank battalions had an even mix of Grants and Shermans. None of the unit records provide exact numbers for which type. They normally just give the number of tanks total that they had for an attack or that were destroyed.

36. Hamilton Howze, "Artillery Tank Support," *Field Artillery Journal* 33 (October 1943): 779.

37. Howze, *Tank Action*, 4. My combat experience in a tank in Iraq in 2003 brought on the same exact feeling described by Howze sixty years earlier.

38. FlaK stands for *Flugabwehr-Kanone* in German.

39. Howe, *"Old Ironsides,"* 245.

40. "Armored Forces," *Intelligence Bulletin* 1, no. 1 (September 1942): 23; "Notes from the North African Theater," *Intelligence Bulletin* 1, no. 4 (December 1942); "New 75mm Antitank Gun," *Intelligence Bulletin* 1, no. 11 (July 1943): 22. Also see both of the following volumes for detailed information on German antitank guns: T. J. Gander, *German Anti-tank Guns* (London: Almark Publications, 1973); Werner Haupt, *German Anti-tank Guns* (West Chester, Pa.: Schiffer Publishing, 1990).

41. Charles Scott, "Organization, Training, and Operations in the Middle East," 21 August 1942, N2815, CARL, 13. Major General Scott was the senior U.S. military observer in the Middle East from March to July 1942.

42. "German Anti-tank Training," no author, n.d., N6137, CARL, 1. I believe that American observers obtained this document from the British army in the spring of 1942, but it was filed poorly at the CARL library.

43. A "hull down position" is a term utilized by tankers to describe a favorable defensive position for a tank. This means that a piece of terrain is blocking the hull of the tank from enemy fire while the turret remains above this terrain feature to engage the enemy. This type of position improves the survivability of the tank and makes it a smaller target for the enemy.

44. Howze, *Tank Action*, 4–5. He also describes this incident in his Senior Officer Oral History Program interview.

45. Ibid., 6.

46. Ibid.

47. *FM 17-10, Armored Force Field Manual: Tactics and Techniques* (Washington, D.C.: U.S. Government Printing Office, 1942), 97. This section suggests that commanders should deploy tank destroyers at the rear of a formation during an attack. Ward's battle plays memorandum indicates this as well. The battle plays complied with some sections of written doctrine, but they usually provided a great more detail than the manual.

48. Ibid., 198.

49. These standing operating procedures (SOP), often mislabeled "standard operating procedures," are common to all tactical units in the modern U.S. Army. They are typically referred to as TACSOPs (tactical SOPs) and are printed up in small pamphlet form for easy distribution in combat environments. Units often review and update these procedures after combat or significant training events.

50. Training Memorandum, Number 127, "Battle Plays," 9 September 1942, MHI, 1. General Ward had this document disseminated throughout his division in September 1942. This left two months to disseminate and practice these plays in addition to any training time that a unit picked up in North Africa. The document includes numerous diagrams of the plays. The author of the document was Colonel C. C. Benson, who would later command a task force after Kasserine Pass and use these same plays.

51. Howe, *"Old Ironsides,"* 215; David Rolf, *The Bloody Road to Tunis: Destruction of the Axis Forces in North Africa, November 1942–May 1943* (London: Greenhill Books, 2001), 269. The cliff that Crosby advanced on was so tenuous that he actually lost one of his tanks when it fell off the cliff down a 30-foot drop. Advancing in armored vehicles under attack is never easy, but it becomes infinitely more difficult when the danger of falling off a cliff becomes part of the equation.

52. Action AT El Guetar [Guettar], April 1st, 9 September 1943, Orlando Ward Papers, MHI; Interview with Captain John McWatters, Orlando Ward Papers, MHI.

53. Training Memorandum, Number 127, "Battle Plays," 9 September 1942, MHI, 1.

54. A Memorandum to the Members of the First Armored Division, 27 February 1943, Orlando Ward Papers, MHI. Ward wrote this memorandum after Kasserine and during a period when the unit was consolidating and retraining. While he does not fault his superiors, Ward encourages his soldiers to integrate all combined arms better, especially infantry and artillery.

55. Howze, *Tank Action*, 7. Both Allied and Axis forces used olive groves for concealment. The trees offered one of the few spots that units could conceal supply dumps or vehicles on a valley floor. While some coniferous forests existed on the slopes of the mountains, the valley floors did not possess forests or vegetation of any significance.

56. Ibid., 9.

57. Ibid.

58. Howze, *A Cavalryman's Story*, 71–75.

59. See Hamilton Howze, SOOHP, MHI, 1–46, for a more detailed background on him.

60. Mark T. Calhoun, "Defeat at Kasserine: American Armor Doctrine, Training, and

Battle Command in Northwest Africa, World War II" (master's thesis, U.S. Army Command and General Staff College, 2003), 1–5. Calhoun concludes that institutional doctrine and poor equipment accounted for the failure at Kasserine.

61. Observer Report, Notes on Recent Operations on the Tunisian Front, 10 March 1943, CARL.

62. *Notes to Accompany History of 3rd BN, 1st AR*, Lieutenant Colonel Cole, RG 407, NARA II.

63. Paul McDonald Robinett, *Armor Command: The Personal Story of a Commander of the 13th Armored Regiment, of the CCB, 1st Armored Division, and of the Armored School during World War II* (Washington, D.C.: McGregor & Werner, 1958), 56.

64. See Howze, *A Cavalryman's Story*, 56–59, for his opinion on Axis air superiority.

65. Hamilton Howze, "Artillery Tank Support," *Field Artillery Journal* 33 (October 1943): 779.

66. Howze, *A Cavalryman's Story*, 72.

67. Howe, *"Old Ironsides,"* 203–204, 250–251; David M. Glantz and Jonathan House, *When Titans Clashed* (Lawrence: University Press of Kansas, 1995), 283.

68. Letter from Gen. Eisenhower to Gen. Marshall (letter summarized into a memorandum for the president of the United States), 11 June 1943, Box 167, Map Room Files, FDR, 1–2.

69. The Germans lost significant numbers of pilots, aircraft, and new tanks in their surrender in North Africa. This is important since pilots require much more training than tankers. Also, the equipment lost strained the German industrial complex even further.

Chapter 7: The Last First Battle, Sicily

1. Training Memorandum Number 40, "Field Exercise 24–25 March 1943," 23 March 1943, William M. Stokes Jr. Collection, Box 1, GCM, Lexington, Virginia, 1–3.

2. Historical Record, Operations, Second Armored Division, Sicily, 22 April to 25 July 1943, R11274.1, CARL, 1–11; Donald Houston, *Hell on Wheels: The 2d Armored Division* (San Rafael, Calif.: Presidio Press, 1977), 153–176.

3. "Lessons from the Sicilian Campaign," Training Memorandum Number 50, 20 November 1943, N2248-4, CARL, 50. This assertion is more confirmation of the effect of firing first discussed in chapter 4.

4. Ibid., 8, 11–12.

5. Houston, *Hell on Wheels*, 176.

6. Carlo D'Este, *World War II in the Mediterranean, 1942–1945* (Chapel Hill, N.C.: Algonquin Books, 1990), 38–47; Rick Atkinson, *The Day of Battle: The War in Sicily and Italy, 1943–1944*, 1st ed., Liberation Trilogy, vol. 2 (New York: Henry Holt, 2007), 1–168; Albert N. Garland and Howard McGaw Smyth, *Sicily and the Surrender of Italy*, United States Army in World War II: Mediterranean Theater of Operations (Washington, D.C.: Center of Military History, United States Army, 1986), 1–18, 58–63, 88–110.

7. "Lessons from the Sicilian Campaign," 66.

8. Ibid., 67.

9. "45th Infantry Division in the Sicilian Campaign," as compiled from G-3 Journal for Period 10 July 1943–22 August 1943, Box 17758, RG 407, NARA II, 2–4.

10. "Lessons from the Sicilian Campaign," 11–12.

11. *Sicily, The U.S. Army Campaigns of World War II*, Center of Military History Publication 72-16, n.d., 19–20.

12. "45th Infantry Division in the Sicilian Campaign," 4–23.

13. Ibid., 9.

14. Terry Allen, "A Summary of the Sicily Campaign, during World War II (10 July–16 Aug 1943) with Special Reference to the Continued Offensive and Night Attacks of the 1st Infantry Division," N3901.5, CARL, 13. The 1st Infantry Division was reinforced by seven additional field artillery battalions for the attacks near Troina, Sicily.

15. Report of Operation, "Participation of 3rd Infantry Division (REINF) in Sicilian Operation, July 10–18, 1943," 10 September 1943, N6528-A, CARL, 41; Ben Franklin Interview, Veterans Oral History Project Center for the Study of War and Society, University of Tennessee—Knoxville, 19 November 2004, 25.

16. Typically, the first thirteen weeks of basic training did not involve any occupation-specific training. Soldiers usually spent another thirteen weeks learning their craft. The additional time also had the effect of instilling discipline and reinforcing lessons learned in the soldier's first thirteen weeks

of training. "Headquarters Armored Force, Unit Mobilization Training Program, 26 Weeks' Schedule," 1 May 1942, Armored Board, RG 337, Box 16, NARA, 1–14.

17. Donald G. Taggart, ed., *History of the Third Infantry Division in World War II* (Nashville, Tenn.: Battery Press, 1990), 447, 479, and 508.

18. Michael King, *William Orlando Darby: A Military Biography* (Hamden, Conn.: Archon Book, 1981), 74–75.

19. Edward B. Kitchens Jr., "The Operations of the 3rd Ranger Infantry Battalion in the Landings at Licata and Subsequent Attack on Porto Empedocle, 10–17 July 1943," WWII Monographs Collection, DRL, 6–7.

20. Field Order #5, HQ RCT 7, 1 July 1943, in Report of Operations, Headquarters RCT 7, 24 July 1943, N15509, CARL, 2. The letter designations are graphic control measures on the operations overlay, and they correspond to towns or terrain features.

21. Ibid., 1–5, annexes II and III.

22. Garland and Smyth, *Sicily and the Surrender of Italy*, 125–129.

23. Participation of 3rd Inf Div (Reinf) in Sicilian Operation, 10–18 July 1943, N6528-A, CARL, 8.

24. Garland and Smyth, *Sicily and the Surrender of Italy*, 129.

25. "Notes on the Planning and Assault Phases of the Sicilian Campaign" by a Military Observer, C.O.H.Q. Bulletin No. Y/1, October 1943, N6530.1, CARL, 26.

26. Ibid., 7–11.

27. "Digest of Reports on Operation 'HUSKY,'" C.O.H.Q. Bulletin Number 62, October 1943, N6530.2, CARL, 2.

28. Ibid., annexure A.

29. Lessons from Operation Torch, Headquarters Second Armored Division, 26 December 1942, N18424.10, CARL, 10; Col. Harry Roper, Report on Observations Made as Observer with Task Force Brushwood, Report of Observers: Mediterranean Theater of Operations, vol. 1, 22 December 1942—23 March 1943, MHI, 10–11.

30. Report of Operations, RCT 7, 24 July 1943, N15509, CARL, 6.

31. *Sicily, The U.S. Army Campaigns of World War II*, 23–24.

32. D'Este, *World War II in the Mediterranean*, 55–56.

33. Clay, *Blood and Sacrifice*, 172,

34. Allen, "A Summary of the Sicily Campaign, During World War II," 5–7; James Scott Wheeler, *The Big Red One: America's Legendary 1st Infantry Division from World War I to Desert Storm*, Modern War Studies (Lawrence: University Press of Kansas, 2007), 232–235.

35. James Lyle, "The Operations of Companies 'A' and 'B' First Ranger Battalion, at Gela, Sicily, 10–11 July 1943," DRL, 6–11, 20–23.

36. King, *Darby*, 77–87. Wheeler, *The Big Red One*, 229–231.

37. In a last-minute decision, Patton decided to parachute the 504th PIR into the Gela bridgehead to assist the II Corps's defense against the German

counterattacks. Poor coordination, continual Luftwaffe attacks, the confusion of the day's counterattacks, and the decision not to land the paratroopers in aircraft in daylight on the airfields created the perfect storm for fratricide. U.S. units and ships shot down twenty-three of their own aircraft and killed 141 U.S. servicemen. See "Report of Airborne Operations, Husky and Bigot," 15 August 1943, N64142, CARL, Section V.

38. Garland and Smyth, *Sicily and the Surrender of Italy*, 150–152.

39. "82d Airborne Division in Sicily and Italy," 3 November 1945, R11960, CARL, 10–11.

40. Distinguished Service Cross Citation, Oak Leaf Cluster, Arthur F. Gorham, General Order Number 33, 23 October 1943, CARL. Gorham was awarded this second DSC after all the after action reports were compiled and analyzed.

41. Distinguished Service Cross Citation, Arthur F. Gorham, General Order Number 16, 18 August 1943, HQ, Seventh Army, CARL; "82d Airborne Division in Sicily and Italy," 1.

42. Garland and Smyth, *Sicily and the Surrender of Italy*, 187–189.

43. "Airborne Assault Operations, 9–14 July 1943," 21 August 1943, N12704, CARL, 1–2; "82d Airborne Division in Sicily and Italy," 9–11.

44. "Report of the 505th Parachute Combat Team in the Landing on Sicily, July 10, 1943," n.d., N12705, CARL, 5–6; "82d Airborne Division in Sicily and Italy," 20.

45. Garland and Smyth, *Sicily and the Surrender of Italy*, 173.

46. "Report of Airborne Operations, Husky and Bigot," 15 August 1943, N6414, CARL, 6–7. This is an observer report compiled by the Fifth Army Airborne Training Center.

47. "Lessons from the Sicilian Campaign," Training Memorandum Number 50, 20 November 1943, N2248–4, CARL, 66.

Conclusion

1. Dwight D. Eisenhower, *Crusade in Europe*, 1st ed. (Garden City, N.Y.: Doubleday, 1948), 179. At the conclusion of HUSKY, Eisenhower was convinced that the Allies had finally mastered interservice (joint) and international cooperation and synchronization.

2. Josiah Bunting III, "Gen. George C. Marshall and the Development of a Professional Military Ethic," *Foreign Policy Research Institute* 16, no. 4 (June 2011): 4.

3. Maj. Gen. E. N. Harmon, "Commanding General's Report on Combat Experience of 1st Armored Division," 13 June 1943, Box 1, Folder 2, James R. Pritchard Collection, Patton Museum, 9.

4. Daily Schedule of All Classes, 4 January 1945, Box 5, Van Zant Collection, Patton Museum.

5. Supervisory Check Sheets, 13 December 1944, Box 5, Van Zant Collection, Patton Museum.

6. The professionalism of the U.S. Army is not in debate since that process began in the 1820s, but it is important to note the impacts of the profession on its officers. Certainly, the regular army met the six conditions of a profession. Allan R. Millett, *Military Professionalism and Officership in America* (Columbus: Mershon Center of the Ohio State University, 1977), 2.

7. Statistics pulled from Battalion Leadership Database created by the author. The database was mainly constructed from operations reports and *The Register of Graduates and Former Cadets of the United States Military Academy, West Point, New York* (West Point, N.Y.: Association of Graduates, 2010).

8. Letter from Drew Boatner to Steven Barry, 21 September 2008.

9. Statistics pulled from Battalion Leadership Database created by the author.

10. Dozens of combat observer reports and lessons-learned documents attest to the soundness of U.S. Army doctrine. For example, Lieutenant Colonel E. A. Russell, executive officer for CCB, said, "Generally speaking, the tactical doctrine laid down by our field manuals has proven to be sound." "Reports on Combat Experience and Battle Lessons for Training Purposes," Memorandum to Commanding General, 1st U.S. Armored Division, 10 June 1943, Box 1, Folder 2, Pritchard Collection, Patton Museum, 1. And Major Michael Popowski, executive officer for the 81st Armored Reconnaissance Battalion, said, "The present tactical doctrine on reconnaissance is sound." "Report on Combat Experience and Battle Lessons for Training Purposes," Memorandum to Commanding General, First U.S. Armored Division, 9 June 1943, Box 1, Folder 2, Pritchard Collection, Patton Museum, 1.

11. "Report of Signal Communications," Ground Observer Report, Army Ground Forces, 8 June 1943, N6446, CARL, 4–6.

12. George F. Hofmann, *Through Mobility We Conquer: The Mechanization of U.S. Cavalry* (Lexington: University Press of Kentucky, 2006), 297–299.

13. Letter from Gen. Eisenhower to Gen. Marshall (letter summarized into a memorandum for the President of the United States), 30 April 1943, Box 167, Map Room Files, FDR, 1.

14. Ibid.

15. George S. Patton, *The Patton Papers, 1940–1945*, 2 vols., comp. and ed. Martin Blumenson (Boston: Houghton Mifflin, 1972–1974), 188.

16. Eisenhower, *Crusade in Europe*, 143.

17. Laurence Robertson, "Combat Experiences with Company 'H' First Armored Regiment, First Armored Division," Box 1, Robertson Papers, USMA, 40.

18. Peter R. Mansoor, *The GI Offensive in Europe: The Triumph of American Infantry Divisions, 1941–1945* (Lawrence: University Press of Kansas, 1999), 73–79.

19. Howze, *A Cavalryman's Story*, 129. I could not locate the book Howze

referred to in this passage of his memoirs. I believe that it must have been a U.S. Army publication that attempted to distill Clausewitz's work into a simplified version. However, the title is ironic in that Clausewitz would not have ever advocated the establishment of a set list of principles of war.

Historiography Essay

1. War World II movies have captured the American imagination continually, even during the war. Some classics are *Saints and Soldiers*, *Before the Glory Fades*, *Battle of the Bulge*, *Hell Is for Heroes*, *Attack*, *The Longest Day*, *A Bridge Too Far*, and *Patton*, to name a few.

2. Marshall and Dupuy offer some initial and interesting conclusions about the war, but this study will not delve into their models nor will it attempt to develop an artificial model to evaluate battalion leadership.

3. Please see the following works that discuss military effectiveness: Charles Bailey, *Faint Praise: American Tanks and Tank Destroyers during World War II* (Hamden, Conn.: Archon, 1983); T. N. Dupuy, *Numbers, Predictions, and War* (Indianapolis, Ind.: Bobbs-Merrill, 1979); T. N. Dupuy, "Mythos or Verity? The Quantified Judgement Model and German Combat Effectiveness," *Military Affairs* 50 (October 1986): 204–210; T. N. Dupuy, "A Response to 'The Wehrmacht Mythos Revisited,'" *Military Affairs* 51 (October 1957): 196–197; John Ellis, *The Sharp End: The Fighting Man in World War II* (New York: Scribner, 1980); Christopher Gabel, *Seek, Strike, and Destroy. U.S. Army Tank Destroyer Doctrine in World War II* (Ft. Leavenworth, Kans : Combat Studies Institute, 1985); Lee Kennett, *G.I.: The American Soldier in World War II* (Norman: University of Oklahoma Press, 1997); Adrian Lewis, *The American Culture of War: The History of U.S. Military Force from World War II to Operation Iraqi Freedom* (New York: Routledge, 2007); S. L. A. Marshall, *Men against Fire* (New York: William Morrow, 1947); and Russell Weigley, *Eisenhower's Lieutenants: The Campaign of France and Germany, 1944–1945* (Bloomington: Indiana University Press, 1981), and *The American Way of War: A History of the United States Military Strategy and Policy* (New York: Macmillan, 1973).

4. Max Hastings, *Overlord: D-Day and the Battle for Normandy* (New York: Vintage Books, 1984), 50–51.

5. Martin van Creveld, *Fighting Power: German and U.S. Army Performance, 1939–1945* (Westport, Conn.: Greenwood Press, 1982); John Ellis, *Brute Force: Allied Strategy and Tactics in the Second World War* (New York: Viking, 1990); and John Keegan, *Six Armies in Normandy: From D-Day to the Liberation of Paris* (New York: Penguin Books, 1994).

6. Ernie Pyle, *Here Is Your War: Story of G.I. Joe* (New York: Henry Holt, 1943), 145.

7. James Tobing, *Ernie Pyle's War: America's Eyewitness to World War II* (New York: Free Press, 1997), 81–93. See Pyle, *Here Is Your War*; Ernie Pyle, *Brave*

Men (New York: Henry Holt, 1944); and Ralph Ingersoll, *The Battle Is the Payoff* (New York: Harcourt, Brace, 1943). Ingersoll's writing is a little less praiseworthy than Pyle's but not by much.

8. The works are numerous, but here are a few: David Glantz and Jonathan House, *When Titans Clashed: How the Red Army Stopped Hitler* (Lawrence: University Press of Kansas, 1995), and *The Battle of Kursk* (Lawrence: University Press of Kansas, 1999); David Glantz, *Stumbling Colossus: The Red Army on the Eve of World War* (Lawrence: University Press of Kansas, 1998).

9. Allan R. Millett, "The United States Armed Forces in the Second World War," in *Military Effectiveness*, vol. 3: *The Second World War*, ed. Allan R. Millett and Williamson Murray (Boston: Allen & Unwin, 1988), 45–89; John S. Brown, *Draftee Division: The 88th Infantry Division in World War II* (Lexington: University Press of Kentucky, 1986); John Sloan Brown, "Colonel Trevor N. Dupuy and the Mythos of Wehrmacht Superiority: A Reconsideration," *Military Affairs* 50 (January 1985): 16–20, and "The Wehrmacht Mythos Revisited: A Challenge for Colonel Trevor N. Dupuy," *Military Affairs* 51 (July 1987): 146–147; and Keith Brown, *When the Odds Were Even: The Vosges Mountains Campaign, October 1944–January 1945* (Novato, Calif.: Presidio Press, 1994).

10. James J. Carafano, *GI Ingenuity: Improvisation, Technology, and Winning World War II* (Westport, Conn.: Praeger Security International, 2006); and James Scott Wheeler, *The Big Red One: America's Legendary 1st Infantry Division from World War I to Desert Storm* (Lawrence: University Press of Kansas, 2007).

11. Unfortunately, the U.S. Army's records are not as well organized or catalogued for the MTO. Also, this theater is viewed as a bit of a sideshow compared to the U.S.-favored cross-channel attack, so it is often overlooked or glossed over by historians. Historians have not put forth much analysis of the U.S. Army in the Pacific for some of the same reasons. Often, the Pacific is viewed as a naval and air fight with the Marine Corps dominating ground combat.

12. David Johnson's *Fast Tanks and Heavy Bombers* asserts a dim view of the U.S. Army's ability to envision future warfare in the interwar period due mostly to bureaucratic gatekeeping by individual branches as well as an education system that failed to produce any great thinkers like Karl von Clausewitz.

13. Michael Doubler, *Closing with the Enemy: How GIs Fought the War in Europe, 1944–1945* (Lawrence: University Press of Kansas, 1994); Russell A. Hart, *Clash of Arms: How the Allies Won in Normandy* (Norman: University of Oklahoma Press, 2001); and Peter R. Mansoor, *The GI Offensive in Europe: The Triumph of American Infantry Divisions, 1941–1945* (Lawrence: University Press of Kansas, 1999). See the following for a more negative view of the U.S. Army: David E. Johnson, *Fast Tanks and Heavy Bombers: Innovation in the US Army, 1917–1945* (Ithaca, N.Y.: Cornell University Press, 1998), and Adrian R. Lewis, *Omaha Beach: A Flawed Victory* (Chapel Hill: University of North Carolina Press, 2001). For a balanced approach on armored force effectiveness, see George F. Hofmann and Donn A. Starry, eds., *Camp Colt to Desert Storm: The*

History of US Armored Forces (Lexington: University Press of Kentucky, 1999), and George F. Hofmann, *Through Mobility We Conquer: The Mechanization of U.S. Cavalry* (Lexington: University Press of Kentucky, 2006).

14. James Russell, *Innovation, Transformation, and War: Counterinsurgency Operations in Anbar and Ninewa, Iraq, 2005–2007* (Stanford, Calif.: Stanford Security Studies, 2011), 1–53. Also see Chad Serena, *A Revolution in Military Adaptation: The US Army in the Iraq War* (Washington, D.C.: Georgetown University Press, 2011), for a very broad study of the U.S. Army.

15. Richard Duncan Downie, *Learning from Conflict: The US Military in Vietnam, El Salvador, and the Drug War* (London: Praeger, 1998), 109–128; Terry Terriff, "US Ideas and Military Change in NATO, 1989–1994," in *The Sources of Military Change: Culture, Politics, and Technology*, ed. Theo Farrell and Terry Terriff (Boulder, Colo.: Lynne Rienner Publishers, 2002), 91–118.

16. For a revisionist and favorable view of the Italian contribution to the Axis defeat, please see James Sadkovich, *The Italian Navy in World War II* (Westport, Conn.: Greenwood Press, 1994), "Of Myths and Men: Rommel and the Italians in North Africa, 1940–42," *International History Review* (1991): 284–313, and "Understanding Defeat: Reappraising Italy's Role in World War II," *Journal of Contemporary History* 24 (1989): 27–61.

17. For adulatory accounts of the Germans, see Ronald Lewin, *Rommel as Military Commander* (New York: McGraw-Hill, 1969); Samuel Mitcham Jr., *Triumphant Fox: Erwin Rommel and the Rise of the Afrika Korps* (New York: Stein and Day, 1984); and Desmond Young, *Rommel the Desert Fox* (New York: Harper, 1950). The more balanced accounts are David Irving, *The Trail of the Fox: The Search for the True Field Marshall Rommel* (London: Wiedenfeld and Nicolson, 1977), and Ronald Lewin, *The Life and Death of the Afrika Corps* (New York: McGraw-Hill, 1977). Works with a British perspective are J. A. Agar-Hamilton and L. C. F. Turner, *The Sidi Rezeg Battles, 1941* (London: Oxford University Press, 1957); Correlli Barnett, *The Desert Generals* (Bloomington: Indiana University Press, 1960); and Michael Carver, *Dilemmas of the Desert War: A New Look at the Libyan Campaign, 1940–1942* (Bloomington: Indiana University Press, 1986).

18. Rich Atkinson, *An Army at Dawn: The War in North Africa, 1942–1943* (New York: Henry Holt, 2002); Martin Blumenson, *Kasserine Pass* (New York: Tower Publications, 1966); Norman Gelb, *Desperate Venture: The Story of Operation Torch* (New York: William Morrow, 1992); Charles E. Heller and William A. Stofft, *America's First Battles, 1776–1965* (Lawrence: University Press of Kansas, 1986); George Howe, *Northwest Africa: Seizing the Initiative in the West* (Washington, D.C.: Center of Military History, 1957); Orr Kelly, *Meeting the Fox: The Allied Invasion of Africa, from Operation Torch to Kasserine Pass to Victory in Tunisia* (New York: J. Wiley, 2002); Kenneth Macksey, *Crucible of Power: The Fight for Tunisia, 1942–1943* (London: Hutchinson, 1969); David Rolf, *The Bloody Road to Tunis: Destruction of the Axis Forces in North Africa, November 1942–May 1943* (Mechanicsburg, Pa.: Stackpole Books, 2001); and Charles

Whiting, *Kasserine: The Anatomy of Slaughter* (New York: Scarborough House, 1998). For the Sicilian and Italian campaign narratives, the best are Rick Atkinson, *The Day of Battle* (New York: Henry Holt, 2007); Carlo D'Este, *Bitter Victory: The Battle for Sicily, July–August 1943* (New York: Harper Perennial, 1991); and Dominick Graham and Shelford Bidwell, *Tug of War: The Battle of Italy, 1943–1945* (New York: St. Martin's Press, 1986).

19. Ralph Bennett, *Ultra and the Mediterranean Strategy, 1941–1945* (London: Hamish Hamilton, 1989); Carlo D'Este, *World War II in the Mediterranean, 1942–1945* (Chapel Hill, N.C.: Algonquin Books, 1990); Michael Howard, *The Mediterranean Strategy in the Second World War* (London: HMSO, 1968); W. G. G. Jackson, *The Battle for North Africa, 1940–1943* (New York: Mason/Charter, 1975); Ronald Lewin, *Ultra Goes to War* (New York: McGraw-Hill, 1978); Geoffrey P. Megargee, *Inside Hitler's High Command* (Lawrence: University Press of Kansas, 2000); and I. S. O. Playfair et al., *The Mediterranean and Middle East*, 5 vols. (London: HMSO, 1954–1973).

20. Edward M. Coffman, *The Regulars, The American Army, 1898–1941* (Cambridge, Mass.: Belknap Press of Harvard University Press, 2004), 418.

21. Ibid.

22. Edward Coffman and Peter Herrly, "The American Regular Army Officer Corps between the World Wars: A Collective Biography," *Armed Forces and Society* 3 (1977): 55–75; Roy K. Flint, "The United States Army on the Pacific Frontier, 1899–1939," in *The American Military and the Far East* (Proceedings of the Ninth Military History Symposium, USAF Academy, 1980); Henry G. Gole, *The Road to Rainbow: Army Planning for Global War, 1934–1940* (Annapolis, Md.: Naval Institute Press, 2003); Brian Linn, *Guardians of Empire: The U.S. Army and the Pacific, 1902–1940* (Chapel Hill: University of North Carolina Press, 1997); Edward Miller, *War Plan Orange: The U.S. Strategy to Defeat Japan, 1897–1945* (Annapolis, Md.: Naval Institute Press, 1991); Allan R. Millett, *The General: Robert L. Bullard and Officership in the United States Army, 1881–1925* (Westport, Conn.: Greenwood Press, 1975); and Lucian Truscott, *The Twilight of the U.S. Cavalry: Life in the Old Army, 1917–1942* (Lawrence: University Press of Kansas, 1989).

Selected Bibliography

Author's Note: This bibliography includes the most relevant resources and publications that I utilized for this book; I have omitted works that I cited only for background information. I also did not include theses and dissertations in the bibliography since I did not solely rely on them for the references and usually paired them with published works or primary resources.

Primary Sources

Air Force Historical Research Agency, Maxwell Air Force Base, Montgomery, Alabama

Bickert, D. Wolfgang. "The Impact of Allied Air Attacks on German Divisions and Other Army Forces in Zones of Combat," 1958.
U.S. Air Force Historical Study No. 30, Ninth Air Force in the Western Desert Campaign to 23 January 1943.
U.S. Air Force Historical Study No. 37, Participation of the Ninth and Twelfth Air Forces in the Sicilian Campaign.
U.S. Air Force Historical Study No. 105, Air Phase of the North African Invasion, November 1944.
U.S. Air Force Historical Study No. 114, The Twelfth Air Force in the North African Winter Campaign, 11 November 1942 to the Reorganization of 13 February 1943, January 1946.

Combined Arms Research Library, Ft. Leavenworth, Kansas

Army Ground Forces Board, "The Tank Destroyer Battalion in Action."
Army Ground Forces Board Reports for North Africa and Sicily.
Army Ground Forces Observer Reports for the Mediterranean Theater of Operations. 15 reports dated from November 1942 to April 1943; 7 reports dated from July to August 1943.
Cavalry Reconnaissance, nos. 1–7, Cavalry School.
Combined Operations Headquarter Bulletins for the Mediterranean Theater of Operations—Operation TORCH—Operation HUSKY.
Comments by LTG Lesley McNair on 2nd Phase, GHQ-Directed Maneuvers.
Eighth Army First Lessons from "Husky."

Operation of II Corps, U.S. Army in Tunisia.
Report on American Airborne Phase of Operation "HUSKY."
Report on Second and Third Army Maneuvers, 1941.
Third Army Maneuvers, May 5–25, 1940.
Unit Reports and Combat Lessons: 1st Armored Division, 2nd Armored Division, 1st Infantry Division, 3rd Infantry Division, 9th Infantry Division, 34th Infantry Division, 45th Infantry Division, 82nd Airborne Division, 1st Ranger Battalion.

Donovan Research Library, Ft. Benning, Georgia

World War II Monographs Collection

Bailey, Leslie W. "The Operations of the 3rd Battalion, 135th Infantry (34th Infantry Division) at Algiers, North Africa, 7 November–10 November 1942 (The Algeria-French Morocco Campaign)."
Carter, Sam. "The Operations of the 1st Battalion, 18th Infantry (1st Division) at El Guettar, Tunisia, 17–25 March 1943."
Davis, Charles L. "The Operations of the Third Platoon, Company 'D,' 13th Armored Regiment (1st Armored Division) in the Vicinity of Matuer, Tunisia, 1 May to 9 May 1943 (Tunisian Campaign)."
Frasher, Ross. "The Operations of the 2nd Battalion, 168th Infantry (34th Infantry Division) at Hill 473 (Djebel El Hara), Northwest of Sidi Nsir, Tunisia, 26 April–28 April 1943."
Heil, George J. "The Operations of Company F, 16th Infantry (1st Infantry Division) at El Guettar, Tunisia, 29 March 1943 (Tunisian Campaign)."
Kron, Jack J. "Operations of the Firing Battery, Battery 'D,' 2nd Bn., 36th F.A., 13th F.A. Brigade (II Corps), in the Fondouk Campaign, 9 March-14 April, 1943 (Tunisian Campaign)."
Lake, Jack F. "The Operations of the 168th Infantry, 2d Battalion (34th Division), Faid Pass, 12–21 February 1943 (Tunisian Campaign)."
Milotta, David E. "The Operations of the First Battalion, 16th Infantry (the First United States Infantry Division) in the Attack on Hill 523, 29–30 April 1943."
Morrissey, Stephen B. "Operations of Company L, 26th Infantry, First Division, in the Night Raid on Maknassy, 17 December 1942 (Tunisia Campaign)."
Ralph, Stephen. "The Operations of the 2nd Battalion, 16th Infantry (1st Infantry Division) in the Invasion of North Africa (Oran, Algeria), 8–11 November, 1942 (The Algeria-French Morocco Campaign)."
Ramsey, David L. "The Operations of the 894th Tank Destroyer Battalion (II Corps) in the Battle of Kasserine Pass, 19–26 February 1943."
Smith, Edwin K. "Operations of the Anti Tank Platoon 2nd BN, 26 INF, 1st DIV, at El Anoor, 8-11 November 1942, Algeria–French Morocco Campaign."

Street, Jack B. "The Operations of the First Ranger Battalion, El Guettar, 21–26 March, 1943 (Tunisian Campaign)."

Tipton, Henry C. "The Operations of the 509th Parachute Infantry Battalion (Separate) in the Invasion of North Africa, 8–13 November 1942 (the North African Campaign)."

Wilson, Lloyd G. "The Operations of the 509th Parachute Battalion (Separate Unit) in North Africa, 8 November 1942–February 1943 (Algerian French Morocco–Tunisian Campaign)."

Franklin D. Roosevelt Presidential Library, Hyde Park, New York

Harry Hopkins Papers.
Franklin D. Roosevelt Presidential Papers.
Map Room Files.

General George Patton Museum Archives, Ft. Knox, Kentucky

Henry E. Gardiner Collection.
George S. Patton Jr. Collection.
James R. Pritchard Collection.
Robert E. Van Zant Collection.

George C. Marshall Foundation, Lexington, Virginia

Memories Project Book Collection.
Personal Papers: John Babicz; Mac Coffman; George Howe; George C. Marshall; Paul Robinett; Beverly M. Read; William M. Stokes Jr.; Lucian Truscott; Walton Walker; Mark S. Watson.
War Department Reports.

Kreitzberg Library, Norwich University Archives and Special Collections, Northfield, Vermont

Personal Papers: James M. Burt; Ernest Harmon; Briard Johnson; George Lucey; Chester Roemmelt; I. D. White Papers.

Library of Congress, Washington, D.C.

George S. Patton Papers.
Paul Robinett Papers.
Theodore Roosevelt Jr. Papers.

Rutgers University, Oral History Archives of World War II, New Brunswick, New Jersey

George Mickett Interview.

Truman Library, Independence, Missouri

Samuel A. Montague oral history interview.

U.S.A. Armor School Research Library, Ft. Knox, Kentucky

"Anti Tank Defense: Weapons and Doctrine." Committee 25, Armored Officers Advanced Course, May 1952.

"Armor Versus Armament." Committee 22, Armored Officers Advanced Course, April 1952.

First Armored Division of the United States Army. Ft. Knox, Ky., 1941.

Moran, Richard. "Communication in a Combat Command."

Sears, Frank S. "Supply Operations in Combat," 1 May 1948.

U.S. Army Center of Military History (CMH), Washington, D.C.

The Armored Force Command and Center. Study No. 27, Historical Section, Army Ground Forces, 1946.

Army Ground Forces Studies.

The Desert Training Center and C-AMA. Study No. 15, Historical Section, Army Ground Forces, 1946.

Ground Forces in the War Army: A Statistical Table. Study No. 3, Historical Section, Army Ground Forces, 1946.

Kasserine Pass Battles. 7 vols. [The CMH developed these volumes to support the 1st Armored Division staff ride to Tunisia in 1993. The volumes contain operational reports, unit histories, personal accounts, excerpts from manuals, and technical data on weapons and vehicles. The CMH pulled most of the material from its own archives and the National Archives. The volumes are mostly repositories of primary sources organized to support a staff ride.]

Mobilization of the Ground Army. Study No. 4, Historical Section, Army Ground Forces, 1946.

Organization and Training of New Ground Combat Elements. Study No. 9, Historical Section, Army Ground Forces, 1946.

The Procurement and Branch Distribution of Officers. Study No. 6, Historical Section, Army Ground Forces, 1946.

Reorganization of Ground Troops for Combat. Study No. 8, Historical Section, Army Ground Forces, 1946.

Sicily, The U.S. Army Campaigns of World War II. Center of Military History Publication 72-16, n.d.

Training in the Ground Army, 1942–1945. Study No. 11, Historical Section, Army.

U.S. Army Military History Institute (MHI), Carlisle, Pennsylvania

Personal Papers: William S. Biddle; Alvan C. Gillem Jr.; Peter C. Hains III; Chester Hanson; Ernest N. Harmon; Hamilton H. Howze; Charles Ivins; Douglas V. Johnson; Sidney Matthew Papers: Russell F. Akers, Ben Crosby, Benjamin A. Dickson, Peter C. Hains III, Ernest N. Harmon; Robert I. Stack; Orlando Ward Papers.

Diaries and Personal Narratives Compiled from Oral History Project: William H. Balzar; Robert G. Bond; Howard F. Kenney; Robert M. Marsh; John Mayo; James Scott Stapel; Ernest A. Wells.

U.S. Army Senior Officer Oral History Program: Paul Adams; Donald Bennett; Andrew Boyle; Arthur S. Collins Jr.; Theodore J. Conway; Harry H. Critz; Michael S. Davison; Welborn Dolvin; Jean Engler; William P. Ennis; Julian Ewell; Paul L. Freeman; Peter C. Hains III; Barksdale Hamlett; Paul D. Harkins; Ben Harrell; John A. Heintges; Robert A. Hewitt; Hamilton Howze; Harold K. Johnson; Robert W. Porter Jr.; Teddy H. Sanford; Richard T. Seitz; George V. Underwood; John K. Waters; Robert J. Wood; William B. Yarborough; Melvin Zais.

Manuscripts:

Camp, T. J., ed. *Tankers in Tunisia*. Armored Replacement Training Center, Fort Knox, Ky., 1943.

Coox, Alvin D., and L. Van Loan Naisawald. "Survey of Allied Tank Casualties." Operations Research Office of Johns Hopkins University, Washington, D.C., 31 March 1951.

Gee, H. G. "The Comparative Performance of German Anti-Tank Weapons during World War II." Army Operational Research Group, Memorandum No. 16, May 1950.

Gee, H. G., and J. M. Murdoch Mitcheson. "Tank Battle Analysis." Military Operational Research Report No. 33.

Hardison, David C. "Data on W.W. II Tank Engagements Involving the U.S. Third and Fourth Armored Divisions." Ballistic Research Laboratories, Memorandum Report No. 798, June 1954.

Howze, Hamilton H. "The Battle of Sidi Bou Zid." Lecture, U.S. Army Cavalry School, n.d.

_____. "Tank Action." 1943.

Ivins, Charles F. "The Monkeys Have No Tails in Zamboanga." May 1974.

Johnson, Douglas V. "Keep 'em Rollin." n.d.

109th Engineer Battalion, Tunisian Campaign History.

Organization and Command Relationships during World War II. Armed Forces Staff College, 17 December 1951.

Report of Operations (Final After Action Report), 12th Army Group. Vol. 11.

Report of Operations, 1st Armored Division, Maknassy, Tunisia, 20 October 1943.

701st Tank Destroyer Battalion: North African Campaign Diary.

Statistical Review, World War II. Army Service Forces, War Department, n.d.

A Survey of Tank Casualties. Department of the Scientific Adviser to the Army Council, Military Operational Research Report, Battle Study No. 82, Report No. 19.

A Survey of Tank Warfare in Europe from D-Day to 12th August 1944. Army Operational Research Group, Memorandum No. 6, May 1952.

TD Combat in Tunisia. U.S. Army Tank Destroyer School, 1944.

Foreign Military Studies:

Burker, Ulrich. "Commitment of the 10th Panzer Division in Tunisia." FMS, D-174. U.S. Army European Command, 1947.

Feige, Richard. "Relationship between Operations and Supply in Africa." FMS, D-124. U.S. Army European Command, 1947.

Kesselring, Albert. "The Events in Tunisia." FMS, D-066. U.S. Army European Command, 1949.

Lang, Rudolf. "Battles of Kampfgruppe Lang in Tunisia." FMS, D-173. U.S. Army European Command, 1947.

______. "Report on the Fighting of Kampfgruppe Lang." Part 2. FMS, D-166. U.S. Army European Command, n.d.

Nehring, Walter. "The First Phase of the Battle in Tunisia." FMS, D-147. U.S. Army European Command, 1947.

Toppe, Alfred. *Desert Warfare*. FMS.

Von Lievenstein, Kurt. "The Drive via Gafsa against Kasserine Pass." FMS, D-124. U.S. Army European Command, 1947.

Werner, Christoph Wilhelm. "The Battles of the Hermann Goring Division in Tunisia from January to 12 May 1943." FMS, D-085. U.S. Army European Command, 1947.

World War II German Intelligence Collection:

U.S. Army Field Manuals

U.S. War Department. *FM 17-10, Armored Force Field Manual: Tactics and Techniques*. Washington, D.C.: U.S. Government Printing Office, 1942.

______. *FM 17-22, Armored Force Field Manual: Reconaissance Battalion*. Washington, D.C.: U.S. Government Printing Office, 1942.

______. *FM 17-30, Armored Force Field Manual: Tank Platoon*. Washington, D.C.: U.S. Government Printing Office, 1942.

______. *FM 17-32, Armored Force Field Manual: The Tank Company, Light and Medium*. Washington, D.C.: U.S. Government Printing Office, 1942.

______. *FM 17-33, Armored Force Field Manual: The Armored Battalion, Light and Medium*. Washington, D.C.: U.S. Government Printing Office, 1942.

______. *FM 21-6, List of Publications for Training, Including Training Films and Film Strips*. Washington, D.C.: U.S. Government Printing Office, 1942.

______. *FM 100-5, Field Service Regulations: Operations*. Washington, D.C.: U.S. Government Printing Office, 1941.

______. *FM 100-20, Command and Employment of Air Power*. Washington, D.C.: U.S. Government Printing Office, 1943.

U.S. Military Academy Library, Special Collections, West Point, New York

Annual Reports of the Superintendent, 1929–1940. *Assembly* alumni magazine, 1943–2008.

Department of Civil and Military Engineering Records, 1931–1941, Series 70.
Howitzer yearbooks, 1929–1940.
Infantry School Mailing List, 1930–1939.
Official Register of the Officers and Cadets, 1929–1940.
The Pointer, 1929–1940.
The Register of Graduates and Former Cadets of the United States Military Academy, West Point, New York. West Point, N.Y.: Association of Graduates, 2008.
The Register of Graduates and Former Cadets of the United States Military Academy, West Point, New York. West Point, N.Y.: Association of Graduates, 2010.
Report on the Army, Biennial Reports of Gen. George C. Marshall, July 1, 1939 to June 30, 1943 to the Secretary of War.
Reports of the Secretary of War to the President, 1930–1939.
United States Corps of Cadets Training Schedules, 1930–1939, Series 111.
War Department Annual Reports, 1929–1941.

Personal Papers: James D. Alger; William O. Darby; John F. Davis; Benjamin A. Dickson; Harry Flint; Floyd Forte; Henry E. Gardiner; Frederick Gilbreath; German Lange; Alex Nininger; Laurence Robertson; William Wedemeyer.

U.S. National Archives, College Park, Maryland

Battle Casualties and Nonbattle Deaths in World War II, Final Report, 7 December 1941–31 December 1946, no date.
RG 156, Ordnance Department.
RG 165, Records of the War Department.
RG 319, Records of the Army Staff General and Special Staffs.
RG 337, Headquarters, Army Ground Forces—Armored Board, Armored School, and Desert Training Center.
RG 407, World War II Operational Reports of the following units: II Corps 1st Armored Division, 1st Infantry Division, 2nd Armored Division, 3rd Infantry Division, 9th Infantry Division, 34th Infantry Division, 45th Infantry Division, 601st, 701st, 894th, and 899th Tank Destroyer Battalions.

University of Tennessee, Veterans Oral History Project, Center for the Study of War and Society, Knoxville, Tennessee

Ben Franklin Interview.
Charles Morton Jr. Interview.
Rodney Quin Nelson Interview.

War Archives (Kriegsarchiv), Austrian State Archives, Vienna, Austria

Friedrich Franek, B/773:8

York County Heritage Trust, York, Pennsylvania
Jacob L. Devers Collection

Published Primary Sources

Bradley, Omar Nelson, and Herman Finkelstein Collection (Library of Congress). *A Soldier's Story.* 1st ed. New York: Holt, 1951.
Cavalry School, United States Army. *Cavalry Combat.* Harrisburg, Pa.: Telegraph Press, 1937.
Eisenhower, Dwight D. *Crusade in Europe.* 1st ed. Garden City, NY.: Doubleday, 1948.
Eisenhower, Dwight D., and Alfred Dupont Chandler. *The Papers of Dwight David Eisenhower: The War Years.* Baltimore: Johns Hopkins University Press, 1970.
Harmon, Ernest Nason, with Milton MacKaye, and William Ross MacKaye. *Combat Commander: Autobiography of a Soldier.* Englewood Cliffs, NJ.: Prentice-Hall, 1970.
Howze, Hamilton H. *A Cavalryman's Story: Memoirs of a Twentieth-Century Army General.* Washington, D.C.: Smithsonian Institution Press, 1996.
Patton, George S. *The Patton Papers.* 2 vols. Comp. and ed. Martin Blumenson. Boston: Houghton Mifflin, 1972–1974.
Patton, George S., Paul D. Harkins, and Beatrice Banning Ayer Patton. *War as I Knew It.* Boston: Houghton Mifflin, 1947.
Robinett, Paul McDonald. *Armor Command: The Personal Story of a Commander of the 13th Armored Regiment, of the CCB, 1st Armored Division, and of the Armored School during World War II.* Washington, D.C.: McGregor & Werner, 1958.
Rommel, Erwin. *Rommel Papers.* Ed. B. H. Liddell-Hart. New York: Harcourt, Brace, 1953.
Truscott, Lucian King. *Command Missions, a Personal Story.* 1st ed. New York: Dutton, 1954.

Secondary Sources

Anders, Leslie. "Four Decades of American Military Journalism." *Military Affairs* 41, no. 2 (April 1977): 62–66.
______. "The Watershed: Forrest Harding's Infantry Journal, 1934–1938." *Military Affairs* 40, no. 1 (February 1976): 12–13.
Atkinson, Rick. *An Army at Dawn: The War in North Africa, 1942–1943.* 1st ed. New York: Henry Holt, 2002.
———. *The Day of Battle: The War in Sicily and Italy, 1943–1944.* 1st ed. Liberation Trilogy, vol. 2. New York: Henry Holt, 2007.
Baily, Charles M. *Faint Praise: American Tanks and Tank Destroyers during World War II.* Hamden, Conn.: Archon Books, 1983.

Behrendt, Hans-Otto. *Rommel's Intelligence in the Desert Campaign, 1941–1943* [Rommels Kenntnis vom Feind im Afrikafeldzug. English.]. London: W. Kimber, 1985.

Bennett, Ralph Francis. *Intelligence Investigations: How Ultra Changed History: Collected Papers of Ralph Bennett*. Cass Series—Studies in Intelligence. London: Frank Cass, 1996.

———. *Ultra and Mediterranean Strategy*. London: Hamish Hamilton, 1989.

Blumenson, Martin. *Kasserine Pass*. Boston: Houghton Mifflin, 1967.

———. *Rommel's Last Victory: The Battle of Kasserine Pass* London: Allen & Unwin, 1968.

Bonn, Keith E. *When the Odds Were Even: The Vosges Mountains Campaign, October 1944–January 1945*. Novato, Calif.: Presidio Press, 1994.

Booth, T. Michael, and Duncan Spencer. "The Airborne's Watery Triumph." In *No End Save Victory: Perspectives on World War II*, ed. Robert Cowley. New York: G. P. Putnam's Sons, 2001.

Borgwardt, Elizabeth. *A New Deal for the World: America's Vision for Human Rights*. Cambridge, Mass.: Belknap Press of Harvard University Press, 2005.

Brown, John Sloan. *Draftee Division: A Study of the 88th Infantry Division, First All Selective Service Division into Combat in World War II*. New York: Presidio Press, 1984.

Bruce, Colin John. *Invaders: British and American Experience of Seaborne Landings, 1939–1945*. Annapolis, Md.: Naval Institute Press, 1999.

Cameron, Robert. *To Fight or Not to Fight?* Fort Leavenworth, Kans.: Combat Studies Institute Press, 2010.

Carafano, James. *GI Ingenuity*. Westport, Conn.: Praeger Security International, 2006.

Center of Military History. *To Bizerte with the II Corps, 23 April–13 May 1943*. CMH Pub. vol. 100-6. Washington, D.C.: Center of Military History, U.S. Army, U.S. G.P.O., 1990.

Citino, Robert Michael. *Death of the Wehrmacht: The German Campaigns of 1942*. Modern War Studies. Lawrence: University Press of Kansas, 2007.

Clay, Steven. *Blood and Sacrifice: The History of the 16th Infantry Regiment from the Civil War through the Gulf War*. Wheaton, Ill.: Cantigny First Division Foundation, 2001.

Coffman, Edward M. *The Regulars: The American Army, 1898–1941*. Cambridge, Mass.: Belknap Press of Harvard University Press, 2004.

Coffman, Edward M., and Peter F. Herrly. "The American Regular Army Officer Corps between the World Wars." *Armed Forces and Society* 4, no. 1 (Fall 1977): 55–73.

Cooper, Belton. *Death Traps*. New York: Ballantine Books, 1998.

Crackel, Theodore J. *West Point: A Bicentennial History*. Lawrence: University Press of Kansas, 2002.

D'Este, Carlo. *Bitter Victory: The Battle for Sicily, July–August 1943*. London: Collins, 1988.

———. *Major Battles and Campaigns: World War II in the Mediterranean, 1942–1945*. Chapel Hill, N.C.: Algonquin Books, 1990.

———. *Patton: A Genius for War*. 1st ed. New York: HarperCollins, 1995.

———. *World War II in the Mediterranean, 1942–1945*. Major Battles and Campaigns vol. 2. Chapel Hill, N.C.: Algonquin Books, 1990.

Dixon, Joe C., ed. *The American Military and the Far East: Proceedings of the Ninth Military History Symposium, United States Air Force Academy, 1–3 October 1980*. Colorado Springs, Colo.: United States Air Force Academy and Office of Air Force History Headquarters, USAF, 1980.

Doubler, Michael D. *Closing with the Enemy: How GIs Fought the War in Europe, 1944–1945*. Modern War Studies. Lawrence: University Press of Kansas, 1994.

Dupuy, Trevor Nevitt. *Numbers, Predictions, and War: Using History to Evaluate Combat Factors and Predict the Outcome of Battles*. Rev. ed. Fairfax, Va.: Hero Books, 1985.

Ellis, John. *Brute Force: Allied Strategy and Tactics in the Second World War*. 1st U.S. ed. New York: Viking, 1990.

———. *The Sharp End of War: The Fighting Man in World War II*. Newton Abbot, U.K.: David & Charles, 1981.

Funk, Arthur Layton. *The Politics of TORCH: The Allied Landings and the Algiers Putsch, 1942*. Lawrence: University Press of Kansas, 1974.

Gabel, Christopher R. *Seek, Strike, and Destroy: U.S. Army Tank Destroyer Doctrine in World War II*. Leavenworth Papers vol. 12. Fort Leavenworth, Kans.: Combat Studies Institute, U.S. Army Command and General Staff College, 1986.

———. *The U.S. Army GHQ Maneuvers of 1941*. New York: Diane, 1992.

Gander, T. J. *German Anti-tank Guns*. London: Almark Publications, 1973.

Garland, Albert N., and Howard McGaw Smyth. *Sicily and the Surrender of Italy*. United States Army in World War II: Mediterranean Theater of Operations. Washington, D.C.: Center of Military History, United States Army, 1986.

Gelb, Norman. *Desperate Venture: The Story of Operation Torch, the Allied Invasion of North Africa*. New York: W. Morrow, 1992.

Germany and the Second World War. Vols. 1, 3, and 4. Oxford: Clarendon Press, 1990.

Gole, Henry G. *The Road to Rainbow: Army Planning for Global War, 1934–1940*. Association of the U.S. Army Book. Annapolis, Md.: Naval Institute Press, 2003.

Graham, Dominick, and Shelford Bidwell. *Tug of War: The Battle for Italy, 1943–1945*. 1st U.S. ed. New York: St. Martin's Press, 1986.

Greenfield, Kent Roberts, Robert R. Palmer, and Bell I. Wiley. *The Organization of Ground Combat Troops*. United States Army in World War II: Army Ground Forces vol. 2-1. Washington, D.C.: Center of Military History, United States Army, 2004.

Hart, Russell. *Clash of Arms: How the Allies Won in Normandy*. Art of War. Boulder, Colo.: Lynne Rienner Publishers, 2001.

Haupt, Werner. *German Anti-tank Guns*. West Chester, Pa. Schiffer Publishing, 1990.

Haupt, Werner, and James K. W. Bingham. *North African Campaign, 1940–1943* [Afrika feldzug. English.]. London: Macdonald, 1969.

Heckmann, Wolf. *Rommel's War in Africa* [Rommels Krieg in Afrika.]. Garden City, N.Y.: Doubleday, 1981.

Heller, Charles E., and William A. Stofft. *America's First Battles, 1776–1965*. Lawrence: University Press of Kansas, 1986.

Hofmann, George F. *Through Mobility We Conquer: The Mechanization of U.S. Cavalry*. Lexington: University Press of Kentucky, 2006.

Hofmann, George F., and Donn A. Starry. *Camp Colt to Desert Storm: The History of U.S. Armored Forces*. Lexington: University Press of Kentucky, 1999.

Houston, Donald. *Hell on Wheels: The 2d Armored Division*. San Rafael, Calif.: Presidio Press, 1977.

Howard, Michael Eliot. *The Mediterranean Strategy in the Second World War*. London: Greenhill Books, 1993.

Howe, George F. *The Battle History of the 1st Armored Division, "Old Ironsides."* Washington, D.C.: Combat Forces Press, 1954.

———. *Northwest Africa: Seizing the Initiative in the West*. United States Army in World War II: Mediterranean Theater of Operations. Washington, D.C.: Office of the Chief of Military History, Dept. of the Army, 1957.

Hoy, Charles J. "The Last Days in Tunisia." *Cavalry Journal* 53 (January–February 1944): 8–12.

_____. "Mechanics of Battlefield Reconnaissance." *Cavalry Journal* 53 (May–June 1944): 24–29.

_____. "Reconnaissance Lessons from Tunisia." *Cavalry Journal* 52 (November–December 1943): 16–21.

Hunnicutt, R. P. *A History of the American Light Tank*. Novato, Calif.: Presidio, 1992.

Irving, David John Cawdell. *The Trail of the Fox*. 1st ed. New York: Dutton, 1977.

Jackson, W. G. F. *The North African Campaign, 1940–43*. London: B. T. Batsford, 1943.

Jacobsen, Hans Adolf, and Jürgen Rohwer. *Decisive Battles of World War II: The German View* [Entscheidungsschlachten des zweiten Weltkrieges.]. 1st U.S. ed. New York: Putnam, 1965.

Johnson, David E. *Fast Tanks and Heavy Bombers: Innovation in the U.S. Army, 1917–1945*. Cornell Studies in Security Affairs. Ithaca, N.Y.: Cornell University Press, 1998.

Johnson, H. K. "Defense along the Abucay Line." *Military Review* 28, no. 11 (February 1949): 43–53.

Kelly, Orr. *Meeting the Fox: The Allied Invasion of Africa, from Operation Torch to Kasserine Pass to Victory in Tunisia*. New York: J. Wiley, 2002.

Kennedy, Paul. "History from the Middle: The Case of the Second World War." *Journal of Military History* 74 (January 2010): 35–51.

Kennett, Lee B. *G.I.: The American Soldier in World War II*. 1st ed. New York: Scribner, 1987.

Killingray, David, and David E. Omissi. *Guardians of Empire: The Armed Forces of the Colonial Powers, c. 1700–1964*. Studies in Imperialism. Manchester, U.K.: Manchester University Press, 1999.

King, Michael. *William Orlando Darby: A Military Biography*. Hamden, Conn.: Archon Book, 1981.

Kretchik, Walter. *US Army Doctrine: From the American Revolution to the War on Terror*. Lawrence: University Press of Kansas, 2011.

Leighton, Richard M., Robert W. Coakley, and Center of Military History. *Global Logistics and Strategy, 1940–1943*. CMH Pub. vol. 1-5. Washington, D.C.: Center of Military History, U.S. Army, 1995.

Lewin, Ronald. *The Life and Death of the Afrika Korps: A Biography*. London: B. T. Batsford, 1977.

———. *Ultra Goes to War: The First Account of World War II's Greatest Secret Based on Official Documents*. New York: McGraw-Hill, 1978.

Linn, Brian McAllister. *The Echo of Battle: The Army's Way of War*. Cambridge, Mass.: Harvard University Press, 2007.

______. *Guardians of Empire: The U.S. Army and the Pacific, 1902–1940*. Chapel Hill: University of North Carolina Press, 1997.

______. "The Long Twilight of the Frontier Army." *Western Historical Quarterly* 27 (Summer 1996): 141–167.

MacDonald, Charles B. *Company Commander*. New York: Bantam Books, 1947.

Macksey, Kenneth. *Crucible of Power: The Fight for Tunisia, 1942–1943*. London: Hutchinson, 1969.

Mansoor, Peter R. *The GI Offensive in Europe: The Triumph of American Infantry Divisions, 1941–1945*. Lawrence: University Press of Kansas, 1999.

Marshall, S. L. A. *Men against Fire: The Problem of Battle Command in Future War*. Washington, D.C.: Combat Forces Press, 1954.

Maycock, Thomas J. "Notes on the Development of AAF Tactical Air Doctrine." *Military Affairs* 14 (Winter 1950): 186–191.

Megargee, Geoffrey P. *Inside Hitler's High Command*. Lawrence: University Press of Kansas, 2000.

Messenger, Charles. *The Tunisian Campaign*. London: Ian Allan, 1982.

Millett, Allan Reed, and Williamson Murray. eds. *Military Effectiveness*. Boston: Allen & Unwin, 1988.

———. *Military Innovation in the Interwar Period*. Cambridge: Cambridge University Press, 1996.

Moran, Charles. *The Landings in North Africa, November 1942*. Washington, D.C.: Naval Historical Center, Dept. of the Navy, 1993.

Mortensen, Daniel R. *A Pattern for Joint Operations: World War II Close Air Sup-*

port North Africa. Washington, D.C.: Office of Air Force History and U.S. Army Center of Military History, 1987.

Morton, Louis. *The Fall of the Philippines*. Washington, D.C.: U.S. Government Printing Office, 1953.

Noble, Dennis L. *The Eagle and The Dragon: The United States Military in China, 1901–1937*. New York: Greenwood Press, 1990.

Oliver, Lunsford E. "In the Mud and Blood of Tunisia." Collier's, 17 April 1943, 10–12.

Ossad, Steven L., and Don R. Marsh. *Major General Maurice Rose: World War II's Greatest Forgotten Commander*. New York: Taylor Trade Publishing, 2003.

Paine, G. H. "Polo in the Philippines." *Infantry Journal* 30 (April 1927): 417–418.

Palmer, Robert R., Bell I. Wiley, William R. Keast. *The Procurement and Training of Ground Combat Troops: The Army Ground Forces*. CMH Pub. vol. 2-2. Washington, D.C.: Center of Military History, U.S. Army, 2003.

Philipsborn, Martin, Jr., and Milton Lehman. "The Untold Story of Kasserine Pass." *Saturday Evening Post*, 14 Feb 1948, 23–28.

Playfair, Ian Stanley Ord. *The Mediterranean and Middle East*. History of the Second World War: United Kingdom Military Series. Rev. ed. London: HMSO, 1974.

Pope, Steven W. "An Army of Athletes: Played Fields, Battlefields, and the American Military Sporting Experience, 1890–1920." *Journal of Military History* 59, no. 3 (July 1995): 435–456.

Porch, Douglas. *The Path to Victory: The Mediterranean Theater in World War II*. 1st ed. New York: Farrar, Straus and Giroux, 2004.

Raff, Edson. *We Jumped to Fight*. New York: Eagle Books, 1944.

Rolf, David. *The Bloody Road to Tunis: Destruction of the Axis Forces in North Africa, November 1942–May 1943*. London: Greenhill Books, 2001.

Rose, Maurice. "Panama's Irregular Cavalry." *Cavalry Journal*, July–August 1935, 26.

Rutherford, Ward. *Kasserine: Baptism of Fire*. Ballantine's Illustrated History of World War II. Battle Book no. 18. New York: Ballantine Books, 1970.

Slocum, A. N., Jr. "The Armored Division in Exploitation." *Military Review* 24 (January 1945): 49–54.

Sorley, Lewis. *Thunderbolt: General Creighton Abrams and the Army of His Times*. New York: Simon and Schuster, 1992.

Steele, Richard W. *The First Offensive 1942: Roosevelt, Marshall and the Making of American Strategy*. Bloomington: Indiana University Press, 1973.

Stoler, Mark A. *Allies in War: Britain and America against the Axis Powers, 1940–1945*. London: Hodder Arnold, 2005.

______. *George C. Marshall: Soldier-Statesman of the American Century*. Boston: Twayne, 1989.

———. *The Politics of the Second Front: American Military Planning and Diplomacy in Coalition Warfare, 1941–1943*. Contributions in Military History vol. 12. Westport, Conn.: Greenwood Press, 1977.

Taggart, Donald G., ed., *History of the Third Infantry Division in World War II*. Nashville, Tenn.: Battery Press, 1990.

Towne, Allen N. *Doctor Danger Forward: A World War II Memoir of a Combat Medical Aidman, First Infantry Division*. Jefferson, N.C.: McFarland, 2000.

Truscott, Lucian King. *The Twilight of the U.S. Cavalry: Life in the Old Army, 1917–1942*. Ed. Lucian K. Truscott III. Modern War Studies. Lawrence: University Press of Kansas, 1989.

Watson, Mark Skinner. *Chief of Staff: Prewar Plans and Preparations*. Washington, D.C.: U.S. Government Printing Office, 1950.

Weigley, Russell F. *History of the United States Army*. New York: Macmillan, 1967.

Wheeler, James Scott. *The Big Red One: America's Legendary 1st Infantry Division from World War I to Desert Storm*. Modern War Studies. Lawrence: University Press of Kansas, 2007.

Whiting, Charles. *Kasserine: First Blood*. New York: Stein and Day, 1984.

Yarborough, William P. *Bail Out over North Africa: America's First Combat Parachute Missions, 1942*. Williamstown, N.J.: Phillips Publications, 1979.

Yeide, Harry. *The Tank Killers: A History of America's World War II Tank Destroyer Force*. Havertown, Pa.: Casemate, 2005.

Index